T0231539

HYPERCONCENTRATED FLOW

INTERNATIONAL ASSOCIATION FOR HYDRAULIC RESEARCH
ASSOCIATION INTERNATIONALE DE RECHERCHES HYDRAULIQUES

IAHR

AIRH   MONOGRAPH SERIES

This monograph was effected under the auspices of the IAHR Section on Fluvial Hydraulics

Hyperconcentrated Flow

ZHAOHUI WAN & ZHAOYIN WANG
Institute of Water Conservancy and Hydroelectric Power Research, Beijing, People's Republic of China

Taylor & Francis
Taylor & Francis Group

LONDON AND NEW YORK

Published by Taylor & Francis
2 Park Square, Milton Park, Abingdon, Oxon, OX14 4RN
270 Madison Ave, New York NY 10016

Transferred to Digital Printing 2007

ISBN 90 5410 166 0

Publisher's Note
The publisher has gone to great lengths to ensure the quality of this reprint but points out that some imperfections in the original may be apparent

Contents

Preface

Hyperconcentrated flow is a peculiar phenomenon in the Yellow River Basin. Significant erosion and siltation associated with hyperconcentrated flood give rise to many problems. On the other hand, irrigation in conjunction with warping by utilizing turbid water at hyperconcentrations greatly enhances agriculture in arid and semi-arid areas. The demand for profound understanding of the mechanism of such flow encourages the study of it. In the past two decades, fruitful results in the study of hyperconcentrated flow were achieved. A feasibility study on utilizing the great potential of sediment-carrying capacity of hyperconcentrated flow in the Yellow River is going on in China. The theory of hyperconcentrated flow can also be applied to hydrotransport, debris flow, sediment release from reservoirs, etc. Furthermore, the development of the theory of hyperconcentrated flow is enriching the knowledge of mechanics of sediment transport.

Early in the fifties, Professor Ning Chien started the study on rheological properties of turbid water. In the sixties, he organized large-scale field surveys of hyperconcentrated flow in rivers. In the following years, he continued active research and organizing hyperconcentrated flow studies. Particularly in the late seventies and in the eighties, he and his research group revealed a series of basic laws of hyperconcentrated flow through a thoughtful arrangement of systematic experiments and field surveys.

As students of Professor Ning Chien, both of us started study on hyperconcentrated flow under his supervision. Thanks to his guidance, remarkable progress has been achieved. Professor Chien passed away in 1986. Now we would like to dedicate this book to him as his memorial.

Prof. P.N. Lin (Lin Bingnan) gave us many valuable instructions during preparation of the manuscript. Prof. H.W. Shen carefully reviewed the draft, gave substantial enlightening advice, and also polished manuscripts. Prof. G. Di Silvio and Prof. M.S. Yalin reviewed the draft. We deeply appreciate their help. Prof. G. Di Silvio, chairman of the IAHR Fluvial Hydraulics Section, recommended to have this monograph published under the auspices of this section. We also acknowledge the help provided by our Chinese colleagues in examining different chapters and giving advice.

Zhaohui Wan & Zhaoyin Wang
Dec. 4, 1992.

CHAPTER 1

Hyperconcentrated flow in nature and in practical application

As a discipline of mechanics of sediment motion, theory of hyperconcentrated flow develops rapidly in the latest two decades, particularly in China. Study on hyperconcentrated flow was started with researches on natural phenomena and engineering problems on the Yellow River. Later on it was found that debris flow, hydrotransport of solid material, etc. are also related to the hyperconcentrated flow.

In an ordinary sediment-laden flow sediment is carried by the flow and sediment has little effect on flow behavior. Therefore such effect can be neglected. In hyperconcentrated flow, however, the existence of large amount of solid particles remarkably influences or changes the fluid properties and flow behavior. In such case the above mentioned influence or change must be considered. In many cases of hyperconcentrated flow sediment together with water, forming a pseudo-one-phase fluid, moves in its entirety and sediment can no longer be considered as material carried by the water.

Whether it is a hyperconcentrated flow can not be simply judged by concentration only. It will be discusssed later that the grain size composition and the mineral content of sediment play very important role. As to the Yellow River where the incoming sediment has similar mineral content and grain size composition, flow with concentration higher than 200 kg/m^3 (or volumetric concentration about 8%) can be considered as a hyperconcentrated one.

1.1 HYPERCONCENTRATED FLOW IN THE MAIN STEM AND TRIBUTARIES OF THE YELLOW RIVER

The Yellow River is notorious for its tremendous amount of sediment. The average annual load is 1.6 billion tons, 80% of which comes from a vast loess plateau in its middle reach. There the land surface, consisting of a chain of undulating hills criss-crossed by thousands upon thousands of gullies (Figure 1.1), is broken up into numerous small watersheds. The erosion- resistance

Figure 1.1. A birds' eye view of the loess plateau (after Yin Hexian).

capacity of the loess area is extremely low because of the loamy texture of the soil and the poor vegetative cover. The loess plateau is an arid/semi-arid area, but there in summer rainstorms may be rather heavy. After saturation, the loose loess deposits with columnar voids readily disintegrate. As a result, summer rainstorms cause severe erosion and hyperconcentrated floods frequently in this area.

In Table 1.1 the maximum and average monthly sediment concentrations of ten main tributaries in the middle reach of the Yellow River are listed (Qi, 1987).

The annual sediment load from these ten tributaries (1.024 billion tons) constitutes 64% of the total load of the Yellow River (1.6 billion tons). Most sediment is transported by hyperconcentrated floods.

Usually the flash hyperconcentrated flood caused by heavy rainstorm lasts a short time, but it always conveys huge amounts of sediment. At some gauging stations concentration higher than 1500 kg/m³ has been recorded. Table 1.2 shows two hyperconcentrated floods recorded at Longmen Gauging Station on the Middle Reach of the Yellow River (Wan & Sheng, 1978). The maximum concentration measured at Longmen Gauging Station is 933 kg/m³ (in 1966).

Sediment carried by hyperconcentrated floods lasting 2-3 days constitutes one-quarter to one-third of the corresponding annual load. Not only large amounts

Table 1.1. Average monthly sediment concentration and maximum recorded sediment concentration of ten main tributaries of the Yellow River.

River	Gauging station	Period of statistics		June	July	August	September	Annual load $(10^6 t)$
Huangfuchuan	Huangfu	1953-1979	S_{av}	411	523	369	216	64.1
			S_{max}	1370	1570	1480	1240	
Gushanchuan	Gaoshiya	1953-1979	S_{av}	327	410	373	178	27.8
			S_{max}	1300	1190	1090	829	
Kuyehe	Wenjiachuan	1953-1979	S_{av}	162	405	319	90.6	135
			S_{max}	1400	1700	1500	970	
Wudinghe	Baijiachuan	1956-1979	S_{av}	12.5	352	323	90.8	106
			S_{max}	1290	1270	1180	958	
Qingjianhe	Yanchuan	1954-1979	S_{av}	384	503	448	105	45.3
			S_{max}	1150	1080	970	881	
Yanshui	Ganguyi	1952-1979	S_{av}	287	454	368	119	54.6
			S_{max}	1200	1190	1033	1070	
Fenhe	Hejin	1943-1979	S_{av}	19	43	59.4	37.6	43.8
			S_{max}	174	386	227	143	
Weihe	Xianyang	1934-1976	S_{av}	37.1	71.4	80.2	28.5	168
			S_{max}	654	588	729	662	
Jinghe	Zhangjiashan	1931-1979	S_{av}	168	349	329	110	286
			S_{max}	906	1430	984	946	
North Luohe	Zhuangtou	1933-1979	S_{av}	121	337	287	58.1	96.8
			S_{max}	987	1150	1190	1340	

S_{av} = average monthly sediment concentration in kg/m^3; S_{max} = maximum recorded sediment concentration in that month in kg/m^3.

Table 1.2. Two hyperconcentrated floods recorded at Longmen Gauging Station.

Flood time	Duration (hrs)	Peak discharge (m^3/s)	Maximum concentration (kg/m^3)	Total load carried by the flood		Thickness of bed erosion (m)
				Amount $(10^6 t)$	Ratio to annual load (%)	
1966 July 18, 9:00 to July 20, 19:00	58	7460	933	453	26.5	7.0
1970 Aug. 2, 0:00 to Aug. 4, 24:00	72	13800	826	497	35.3	8.8

of sediment, but also rapid and severe erosion or deposition are asssociated with hyperconcentrated floods. As seen from Table 1.2, the river bed at Longmen Gauging Station was eroded about nine meters in a very short period (72 hours). Such strong erosion takes its own peculiar form, called 'ripping up the bottom' by local habitants. This phenomenon will be later discussed in detail.

However, in most cases hyperconcentrated flow results in severe siltation. At some gauging stations on small tributaries, the entire river stops moving during the recession of a hyperconcentrated flood, when the discharge becomes smaller but the concentration is still high. The river stops moving for a while, then flows

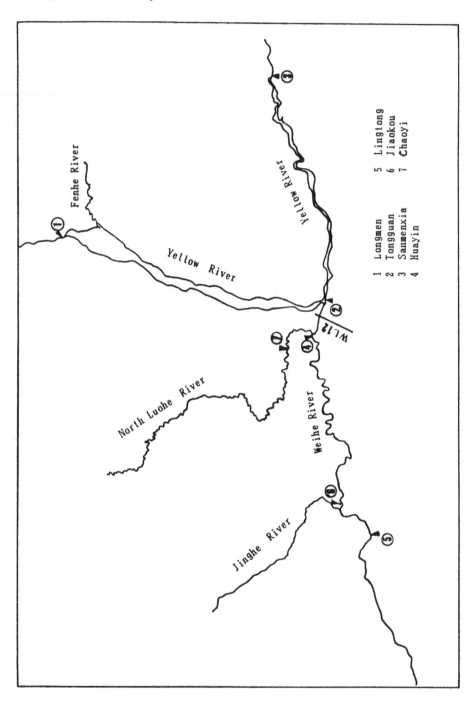

Figure 1.2. Plane view of the middle reach of the Yellow River.

again... Such unstable flow may last for a rather long time and is locally known as 'clogging of the river'. Laboratory studies on such phenomenon (Wan et al., 1979) will be discussed later.

Large amounts of sediment and associated serious erosion and siltation inevitably cause a series of problems in agriculture and industry.

One hundred-three flood events occuring in the Lower Yellow River over a 19-year period have been carefully analysed (Qian et al., 1987). Of these, 13 hyperconcentrated floods contributed to 60% of the total deposition caused by the 103 floods. The average aggradation intensity was as high as 31×10^6 tons per day, much higher than the aggradation intensity caused by other types of floods. Moreover, because such flash floods seldom overflow the flood plain, most of the deposition occurred in the main channel of the river. Consequently, the water surface profile along the river rises rapidly; therefore, floods of this type are most disadvantageous to the flood control of the Lower Yellow River.

Hyperconcentrated floods usually occur in the middle reach of the Yellow River and its tributaries. Weihe River is one of the main tributaries (Figure 1.2 and Table 1.1).

After the completion of Sanmenxia Reservoir in the middle reach of the Yellow River in September, 1960, serious siltation occurred and the backwater deposits extended upstream, seriously endangering the Weihe valley area, which is an important agricultural and industrial base in northwest China. In normal years along the Weihe River the end of the backwater deposits moves upstream. But whenever hyperconcentrated flood with large discharge passes through, the end of the backwater deposits moves downstream due to the intensive erosion along the main channel. Figure 1.3 shows the variation with time of the end of the deposits along the Weihe River. In the figure L is the distance between the end of backwater siltation and Tongguan Guaging Station, the confluence of the Weihe River and the Yellow River (Figure 1.2). In 1964 and 1966 the end moved far downstream, as the result of intensive erosion caused by hyperconcentrated floods. A similar situation also occurs along the stem of the Yellow River upstream from Tongguan. Figure 1.4 shows that after the passage of the second hyperconcentrated flood listed in Table 1.2, the main channel along the upper part of the reach was obviously degraded. In the meantime the flood plain was aggraded.

Intensive erosion causes great trouble for diversion works. In 1977 when a hyperconcentrated flood with large discharge passed through, intensive erosion in the form of 'ripping up the bottom' occurred at Lingtong, where a pumping station with a capacity of 40 m^3/s is located. The local river bed as well as the water level was lowered by 2 m after this event. The bottom of the inlet of the pumping station was above the water surface, so that water could no longer be pumped. An irrigated area of 84 000 hectares suffered from drought for a long time in the summer. A similar situation also occurred in the stem of the Yellow River at Jiamakou, not far upstream from Tongguan.

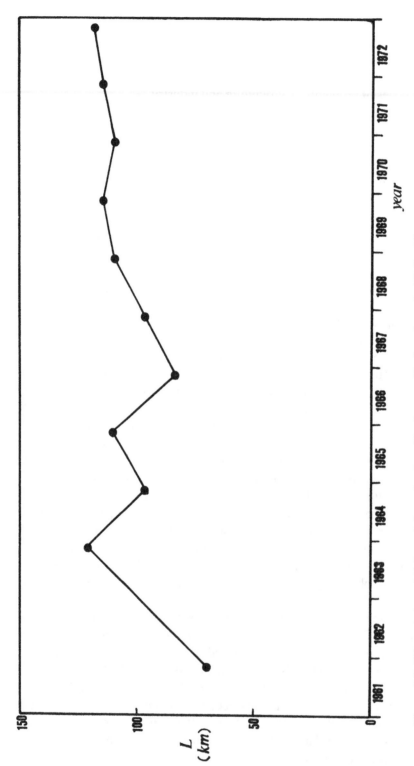

Figure 1.3. The variation with time of the end-point of backwater siltation along the Weihe River.

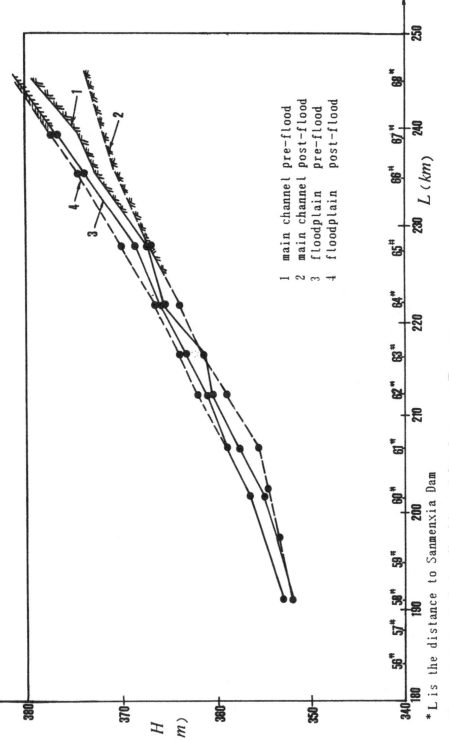

1 main channel pre-flood
2 main channel post-flood
3 floodplain pre-flood
4 floodplain post-flood

*L is the distance to Sanmenxia Dam

Figure 1.4. The longitudinal profile of the reach from Longmen to Tongguan.

The most striking fact was the unusual change of water stage and the rapid, vigorous river deformation associated with the hyperconcentrated flood in the Lower Yellow River in 1977. The variation of the water stage was contrary to that previously predicted by routine method. The water level dropped during the rising limb of the flood and the drop was followed by a rapid rise in the water level (2.84 m in 1.5 hours). Besides, the hyperconcentrated flood changed channel shape greatly. During the passage of hyperconcentrated flood, the wide, shallow, braided river could be transformed into a narrow, deep, meandering river with a single channel. Correspondingly, the velocity in the river channel was so high that many spur dikes and levees were undermined. It became the most critical situation in flood- control since the establishment of the PRC in 1949, and will later be discussed in greater detail.

1.2 HYPERCONCENTRATED FLOW IN RESERVOIRS

Hyperconcentrated flow can pass through a reservoir without severe siltation and can be released from a reservoir if bottom sluices are installed and opened. In this way serious reservoir sedimentation can be avoided or alleviated. Examples of hyperconcentrated flow passing through Sanmenxia reservoir are listed in Table 1.3.

The elevation of the bottom sluices is 280 m above the sea level, and the water depth in front of the dam was 33-37 m during the floods. The slope of the 40 km reach in front of the dam is only 0.0001, or even less. Under such conditions, nearly all the incoming sediment or even more in a few cases was released from the reservoir. In the latter case, some sediment was eroded from the bed in the reach just downstream from Tongguan Gauging Station (Qian et al., 1979).

Heisonglin Reservoir in Shaanxi Province is a small reservoir with a storage capacity of 8.6×10^6 m^3. It is located in a loess plateau region and the concentration of incoming floods is high. The average releasing rate, i.e. the ratio of the amount of sediment released from the reservoir in the form of density current to the total

Table 1.3. Examples of hyperconcentrated flood passing through Sanmenxia Reservoir in 1977.

Time	Tongguan Gauging Station (inlet of the reservoir)			Maximum concentration at the outlet of reservoir (kg/m^3)	Maximum stage in front of the dam (m)	Corresponding storage capacity (10^6 m^3)	Releasing rate (%)*
	Peak discharge (m^3/s)	Maximum concentration (kg/m^3)	Average concentration (kg/m^3)				
July 6	13 600	616	367	589	317	400	91
August 3	12 000	238	178	320	313	190	115
August 6	15 400	911	276	911	315	290	113

*Releasing rate = released sediment/incoming sediment.

amount of incoming sediment, reaches 65% (Xia & Ren, 1980). Bajiazui Reservoir is a large reservoir with a total capacity of 525×10^6 m³. The concentration of incoming flow there is even higher, reaching 573 kg/m³. If the main channel is preserved and the stage in front of the dam is not very high, the releasing rate, which is the ratio between the released sediment and the incoming sediment, may reach 100%. Even under unfavorable conditons, the releasing rate of hyperconcentated density current is still higher than that of an ordinary density current (Jiao, 1989).

Hyperconcentrated density current has a high releasing rate because the density of a hyperconcentrated flow is much higher than that of clear water, so a hyperconcentrated flow can easily plunge to the bottom of a reservoir when it enters the backwater region. Due to its large density, the hyperconcentrated density current also has a much higher velocity than an ordinary density current. Due to its large viscosity, the turbulence in a hyperconcentrated density current is very weak. Therefore, the mixing of a hyperconcentrated density current and clear water at the interface is correspondingly weakened. Besides, particles settle much more slowly in hyperconcentrated flow than in an ordinary density current. As a result, the mixing and deterioration of hyperconcentrated density current along its course is much weaker and slower than that of an ordinary density current. Consequently, more sediment can be transported to the dam site.

Another important factor is the formation of subreservoir in front of the dam. That is, underneath the upper clear water turbid water with hyperconcentration exists. Due to the extremely low settling velocity, once the hyperconcentrated density current arrives at the dam and cannot be entirely released because of

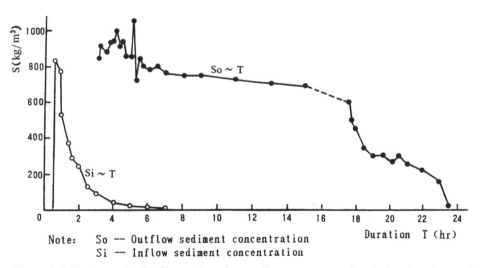

Note: So — Outflow sediment concentration
      Si — Inflow sediment concentration

Figure 1.5. Hydrograph of inflow and outflow sediment concentration during the release of muddy water from the underlying subreservoir, Hengshan reservoir.

Table 1.4. Hyperconcentrated flow released from an underlying muddy water subreservoir.

Date (m.d.y.)	Inflow Smax (kg/m³)	TWi (hr)	Outflow Smax (kg/m³)	TSo (hr)	TSo/TWi	So/Si (%)
6.25.73	469	12.5	1220	206	16.5	48
7.7.75	649	20.5	1240	59	2.9	85
7.2.72	325	12.3	758	42	3.3	53
7.15.76	672	14.8	1015	31	2.1	50
8.13.79	344	12.1	1200	174.5	14.4	76
8.15.80	462	112.6	1010	164.5	1.46	52
7.24.81	833	3.4	1015	21.9	6.4	101
8.13.84	394	2.6	586	12.6	4.8	37.6
8.22.84	356	3.6	1120	38.7	10.5	16.1
7.23.76	407	12.0	851	20.0	1.67	75.2

Note: TWi = duration of incoming flood; TSo = duration of sediment sluicing; So/Si = ratio of sediment load outflow to inflow.

limited discharge capacity, a subreservoir of hyperconcentrated fluid is formed underneath the clear water. Sediment particles in the subreservoir settle at extremely low velocity, perhaps, several centimeters per hour, and the turbid water remains fluid for a rather long time. Within this period the turbid water can be consistently released if the bottom sluices remain open. Figure 1.5 shows an example taken from Hengshan Reservoir in which S_i and S_o are the concentrations of the inflow and the outflow of the reservoir, respectively (Guo et al., 1985). The hyperconcentrated outflow lasted for about twenty hours with two-hours of incoming hyperconcentrated flood. More examples from the same reservoir are listed in Table 1.4.

Notice that in all these cases, the maximum outflow concentration is higher than the maximum inflow concentration. An explanation is that sediment particles in the underlying subreservoir settled, and the muddy water was condensed but remained fluid.

Hyperconcentrated flow can also be formed by emptying a deposited reservoir, when sediment is obtained from retrogressive scour and the lateral slippage of flood plain deposits during reservoir drawdown. The outflow concentration can be rather high. The maximum recorded concentration in Hengshan reservoir is 1320 kg/m³.

If floods occur during the reservoir-emptying period and the dam has sufficient discharge capacity, intensive erosion may be caused by floods. Figure 1.6 depicts a flood recorded in the Hengshan Reservoir. The reservoir has been emptied when a flood with a peak discharge of 20 m³/s arrived. Due to the combined action of retrogressive and longitudinal erosion, hyperconcentrated flow with an average concentration of 666 kg/m³, that was much higher than the average inflow concentration of 212 kg/m³, was formed. Two hundred-seven thousand tons of sediment were sluiced out of reservoir by the flood. In Figure 1.6, S_i and S_o are the

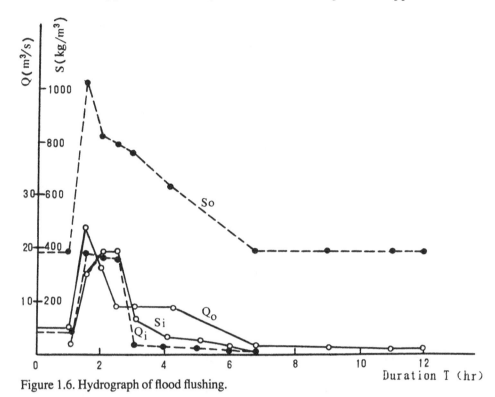

Figure 1.6. Hydrograph of flood flushing.

concentration of the inflow and the outflow of the reservoir, respectively, and Q_i and Q_o are the corresponding discharges.

1.3 HYPERCONCENTRATED DENSITY CURRENT IN RIVERS

Due to its large density and the corresponding large difference in density between it and clear water, a hyperconcentrated flow is liable to form a density current in reservoirs, as well as in rivers. Density current in rivers has been observed at some confluences.

1.3.1 *Density current at the confluence of the Weihe River and the Yellow River (Wan & Niu, 1989)*

The Weihe River converges with the Yellow River at Tongguan. Tongguan Gauging Station is located just downstream from the confluence, as shown in Figure 1.2. Whenever hyperconcentrated flood from the Weihe River pours itself into the Yellow River and the latter is not flooding and relatively clear, the turbid flow of Weihe River immediately plunges underneath the relatively clear water of the Yellow River. A mass of debris, driftwood and leaves accumulating on the water surface along the front of plunging clearly shows the plunging of turbid flow of Weihe River. This phenomenon is quite similar to that associated with the plunging of a density current in a reservoir. In the meantime, at Tongguan Gauging Station a distinct interface marked by abrupt change in both concentration and velocity can be detected by field survey. The plunging of the turbid flow of Weihe River can be proved by comparing the oncoming discharge of Weihe River and the part discharge of the lower turbid layer at Tongguan Gauging Station. Quite often vigorous erosion at Tongguan is associated with the hyperconcentrated density current. All these will be discussed in detail in Chapter 10.

1.3.2 *Density current at the confluence of North Luohe River and Weihe River*

Not far from the confluence of the Weihe River and the Yellow River is the North Luohe River and Weihe River (Figure 1.2). Density current also occurs in this reach (Zeng et al., 1986).

Provided the discharge of the Yellow River is large and that of the Weihe River is small, the lower reach of the Weihe River is influenced by a backwater effect, and the slope of the water surface is very gentle or even reversed. In such situation, a hyperconcentrtated flood coming from the North Luohe River will plunge to the bottom of the Weihe River near the confluence. The turbid water moves both downstream and upstream along the Weihe River, and rapid siltation simultaneously takes place. As an example, velocity profiles taken at Huayin

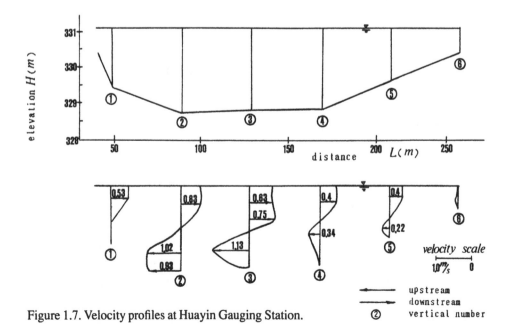

Figure 1.7. Velocity profiles at Huayin Gauging Station.

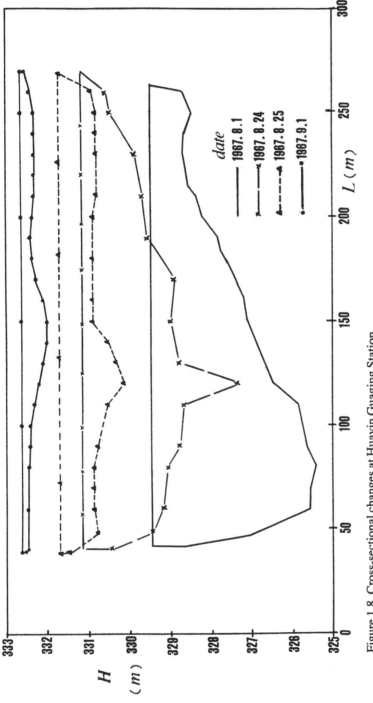

Figure 1.8. Cross-sectional changes at Huayin Guaging Station.

Gauging Station, which is located just upstream of the confluence (Figure 1.2), are shown in Figure 1.7. The upper layer with sediment concentration 28.7 kg/m³ moved downstream and the lower layer with sediment concentration 774 kg/m³ moved upstream at a velocity of more than 1 m/s. This reach was seriously silted during that period. The changes in the cross section at Huayin Gauging Station are shown in Figure 1.8.

1.4 HYPERCONCENTRATED TURBIDITY CURRENT

Density current occurs along a sloping ocean bottom provided liquid adjacent to the bottom contains suspended sediment that causes the average density of the mixture to be greater than the density of the surrounding clear water. Such flow is called turbidity current by geologists. Turbidity current may be initiated by a turbid river entering the sea, by wave action, or by earthquake-induced mud slump. Earthquake-induced mud slump may be of large-scale and develops into a huge turbidity current at extremely high concentrations. It is also a kind of hyperconcentrated flow.

A huge turbidity current was initiated by an earthquake which occurred in the Grand Banks region off Newfoundland in 1929. The progress of the current was measured by the orderly breaking of submarine cables. This phenomenon has been well explained by a series of papers (Heezen & Ewing, 1952; Plapp & Mitchell, 1966; Bagnold, 1962).

Six cables on the continental slope in the epicentral area broke first. Another five cables broke in order indicating the progress of the current down the slope and out onto the ocean bottom. The last cable, located 480 km from the epicenter,

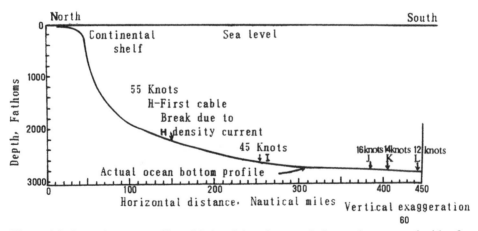

Figure 1.9. Ocean bottom profile, cable break locations, and observed current velocities for the 1929 Grand Banks turbidity current, after Heezen & Ewing (1952).

broke 13 hours and 17 minutes later. The local velocity of the turbidity current could then be estimated by correlating distance and time of breakage. Local velocities and the ocean bottom profile are shown in Figure 1.9 (after Heezen & Ewing, 1952). According to Plapp & Mitchell's (1960) analysis, the thickness of the turbidity current was 300-400 m and the average velocity was over 20 m/s on continental slopes of 6×10^{-3} and 9.8 m/s on the ocean floor where the slope was 10^{-3}.

Despite the great difference in their sizes, turbidity flow and density current in reservoirs share some general characteristics and mechanism.

1.5 HYPERCONCENTRATED FLOW IN CANALS (WAN & XU, 1984)

Most tributaries in the middle reach of the Yellow River carry heavy sediment loads, and hyperconcentrated floods occur quite often. In the past, flows with concentrations higher than 167 kg/m^3 were not allowed to be diverted into irrigation districts because of concern over canal siltation. Irrigation districts suffered from drought even though hyperconcentrated floods passed by their diversion works. In the 1960's in the Luohui Irrigation District people succeeded in raising the concentration limitation for irrigation. Since 1974, field measurement and corresponding laboratory studies have been carried out. The results of these studies indicated that the transport of sediment by hyperconcentrated flow does not require high flow velocity or high flow intensity. In most canals, flow with a velocity of about 1 m/s is enough to carry heavy sediment loads without serious siltation. Experience has been accumulated and a series of rules of thumb for canal design and operation have been worked out. Concentration limitations for irrigation have been abolished and hyperconcentrated flow has been conveyed to most parts of the Luohui Irrigation District. Irrigation by using turbid water at high concentration is called hyperconcentrated irrigation. In 1977, flow at a concentration of 964 kg/m^3 was conveyed through 50 km of canals. The experience and knowledge of hyperconcentrated irrigation obtained in the Luohui Irrigation District is now referred to by other irrigation districts. By the way of hyperconcentrated irrigation/warping not only water, but also nutrient-rich sediments have been utilized as resources. Based on statistics from 1969 to 1985, 2.06×10^8 m^3 of muddy water (concentration higher than 167 kg/m^3), which amounts to 20% of the total water diverted in summer, and 1.06×10^8t of sediment, which amounts to 9.8% of the sediment load of the North Luohe River, were diverted and utilized. On average, 5000 hectares of land were irrigated by muddy water every year, and 3720 hectares of alkaline-saline land have been improved by warping. The gross output value of the land increased by about $10 million due to the use of hyperconcentrated irrigation/warping.

1.6 DEBRIS FLOW

Debris flow is widespread throughout the world, occurring in Japan, Rusland, the United States, China, etc. In the southeastern part of Tibet, the western and northeastern parts of Yunnan Province and the mountainous area of west Sichun Province debris flow occurs quite often. In the Xiaojiang River Basin (Yunan Province), with a total area of 3120 km^2, 500-1000 episodes of debris flow take place every year (Li, 1980). A well equipped experimental station has been established there for the systematic observation and measurement of debris flow (Kang, 1990).

Debris flow is a kind of hyperconcentrated flow. It carries large amounts of granular particles with wide size composition, from large stones to clay particles, and its density may reach 1.9-2.2 g/cm^3. The velocity of debris flow can be rather high. The maximun recorded velocity in China is 13.4 m/s (Zhang & Yuan, 1980). Hence it is a powerful destructive force and threatens railways, highways, lives and the property of local citizens. Volcanic debris flows and other hyperconcentrated flows resulting from them have been observed and studied by Scott & Dinehart (1985). Volcanic debris flows are named as lahars. And the hyperconcentrated streamflow following lahar is named as lahar-runout flow. Lahars are formed in the following ways:

1. By the bulking of lake-breakout flood surges with eroded alluvium;
2. From flood surges produced from snowmelt by hot lithic pyroclastic;
3. From material catastrophically ejected and mixed with water of hydrothermal and glacial or snowmelt origin.

Recorded velocities of lahar-runout flows range from 4 to 7 m/s, which are substantially higher than the common streamflow velocities. The concentration of lahar-runout flows reaches 530-1590 kg/m^3 (20-60% in volume).

It will later be pointed out that there are some similarities in composition size, transport mechanism, and fluvial processes between lahar-runout flow and hyperconcentrated flow in river. Debris flow will be discussed in detail in Chapter 11.

1.7 HYDROTRANSPORT AND DENSECOAL HYDROTRANSPORT

Pipe hydrotransport is a widely adopted form of hydraulic transport. The maximum diameter of existing pipe systems in the world (up to 1990) is 510 mm, and the maximum length is 400 km. The maximum transport capacity amounts to 1.2×10^7 t/y. One of the tendencies of development is to utilize hydrotransport at hyperconcentration. In many hydrotransport systems, the concentration by weight is over 50%. In this way, energy and water can sometimes be saved.

Densecoal is a suspension consisting of coal, water and additives which behaves practically in the same way as oil (Klose & Kunst, 1985). The suspension can be directly burnt as fuel, i.e. without the need for dewatering. Densecoal can be transported by train, ship or pipeline. Densecoal hydrotransport is also a kind of hyperconcentrated flow. Results of studies on rheological properties and mechanics of hyperconcentrated flow in rivers can be referred for densecoal hydrotransport. There are some astonishing similarities between the size composition of hydrotransported material and that of sediment in hyperconcentrated flow in rivers, and many experiments on hyperconcentrated flow have been carried out in pipes or in closed conduits.

1.8 SUMMARY CONCLUSION

Lots of hyperconcentrated flow phenomena occur in nature, agriculture and industry. Some of them are closely related to the development of construction. These phenomena attract the attention of research workers and engineers, and promote the study on hyperconcentrated flow. Some studies have brought people benefits. It can be expected that deep understanding of the mechanism of hyperconcentrated flow will substantially favor the development of agriculture and industry. There are a variety of types of hyperconcentrated flow. Each type of hyperconcentrated flow has its own peculiarities, and together they also have some commonalities. The exploration of the common laws governing various kinds of hyperconcentrated flow will promote the development of sediment transport theory and bring people great advantages.

REFERENCES

Bagnold, R.A. 1962. Auto-suspension of transported sediment: turbidity currents. *Proc. Royal. Soc. London, Ser. A*, Vol. 265(1322): 314-319.

Guo, Z., B. Zhou, L. Ling & D. Li 1985. The hyperconcentrated flow and its related problems in operation at Hengshan Reservoir. *Proc. International Workshop on Flow at Hyperconcentrations of Sediment*: 3-1.

Heezen, B.C. & M. Ewing 1952. Turbidity currents and submarine slumps, and the 1929 Grand Banks earthquarkes. *Amer. J. Sci.*, Vol. 250: 849-873.

Jiao, E. 1989. Study on hyperconcentrated flow in Bajiazui Reservoir, *Report of Institute of Hydraulic Research, Yellow River Conservancy Commission* (in Chinese).

Kang, Z. 1990. Motion characteristics of debris flow at Jiangjia Gully, Yunnan Province, China, Circular No.3. *Publication of Internatinal Research and Training Centre on Erosion and Sedimentation*, pp. 38.

Klose, R & W.D. Kunst, 1985. Densecoal – Densephase flow behaviour of Datong, Fugu and Shenmu coal and densecoal combustion. *Proc. of International Workshop on Flow at Hyperconcentrations of Sediment, Publication of International Research and Training Centre on Erosion and Sedimentation*: 4-2.

Li, J. 1980. Debris flow in Xiaojiang River basin, Yunnan Province (in Chinese). *Proc. Debris Flow by Chengdu Geography Institute*: 34-42.

Plapp, J.E. & J.P. Mitchell 1960. A Hydrodynamic theory of turbidity currents. *J. Geophys. Res.*, Vol. 65(3): 983-992.

Qi, P. 1987. Conveying hyperconcentrated floods through narrow-deep channel into sea is the main measure of solving problems relating to sediment of the Yellow River (in Chinese). *Report of Institute of Hydraulic Research, Yellow River Conservancy Commision.*

Qian, N. (Ning Chien), Z. Wan & Y. Qian 1979. The flow with heavy sediment concentration in the Yellow River Basin (in Chinese). *J. Qinghua University*, Vol.19(2): 1-17.

Qian, N. (Ning Chien), K. Wang, L. Yan & R. Fu 1980. The source of coarse sediment in the middle reaches of the Yellow River and its effect on the siltation of the Lower Yellow River (in Chinese). *Proc. of The International Symposium on River Sedimentation*: 53-62.

Scott, K.M. & R.L. Dinehart 1985. Sediment transport and deposit characteristics of hyperconcentrated streamflow evolved from lahars at mount St. Helens, Washington. *Proc. International Workshop on Flow at Hyperconcentrations of Sediment, Publication of International Research and Training Centre on Erosion and Sedimentation*: 3-2.

Wan, Z. & Z. Niu 1989. Hyperconcentrated density current in rivers. *Proc. of 23th Congress of International Association for Hydraulic Research.*

Wan, Z. & S. Sheng 1978. Hyperconcentrated flow on the Yellow River and its tributaries (in Chinese). *Selected Papers of the Symposium on Sediment Problems on the Yellow River*, Vol. 1(2): 141-158.

Wan, Z. & Y. Xu 1984. The utilization of hyperconcentrated flow and its mechanism. *Proc. 4th Congress APD, IAHR*: 1791-1808.

Wan, Z., Y. Qian, W. Yang & W. Zhao 1979. Laboratory study on hyperconcentrated flow (in Chinese). *People's Yellow River*, No.1, 1979: 53-65.

Xia, M. & Z. Ren 1980. Methods of sluicing sediment from Heisonglin Reservoir and its utilization downstream (in Chinese). *Proc. of The International Symposium on River Sedimentation*, Vol.2: 717-726.

Zeng, Q., W. Zhou & X. Yang 1986. Development of sedimentation for the Weihe River and its relations to Tongguan Constraint and intrusion of flood of the Yellow River (in Chinese). *J. of Sediment Research*, No.3: 13-28.

Zhang, S. & J. Yuan 1980. Impulsive force of debris flow and its measurement (in Chinese). *Proc. Debris Flow by Chengdu Geography Institute*: 137-142.

CHAPTER 2

Basic patterns of motion of hyperconcentrated flow

2.1 DIFFERENT FORMS OF GRAIN MOVEMENT

Although phenomenological and empirical approaches are widely used in the investigation of hyperconcentrated flow, and the results of these approaches are rather encouraging, a complete understanding of hyperconcentrated flow also requires further study on the mechanism involved. Particularly, it is necessary to consider the force by which solid grains are supported in a flowing mixture.

In general, solid particles have larger specific weight than the liquid phase, and tend to settle downward. To maintain the movement of solid grains in flow, a force is needed to balance the submerged weight of solid particles and prevent them from depositing. According to origins of the forces, solid particles carried by a flow can be classified as bed load, suspended load and neutrally buoyant load.

2.1.1 *Bed load*

Bed load refers to the solid grains whose submerged weight (= weight of grains minus buoyancy force) is supported by dispersive force or contact force. When the average shear stress on the bed of an alluvial channel exceeds a critical tractive stress for the bed material, solid particles on the bed statistically may begin to move in the direction of flow. They move in different ways depending on flow conditions, ratio of the density of the fluid and that of particles, and size of particles. One mode of movement of particles is by rolling and sliding on the bed. Sediment transported in this way, whose submerged weight is supported by contact force, is known as contact load. A second mode of bed load movement is by hopping or bouncing along the bed. Thus for some time the particle loses contact with the bed. Material transported in this way is supported by dispersive force and is known as saltation load. Saltation load is an important mode of transport in case of noncohesive material of relatively high fall velocity, such as sand in air and gravels in water. In a few cases, such as in debris flow with less clay material and hydrotransport of cohesionless material in pipelines, such mode of sediment transport may extend into the whole flow depth. These will be discussed in Chapter 7.

The concept of dispersive force was advanced by Bagnold (1954, 1956). He studied collision between solid particles in a hyperconcentrated flow and proposed that the collision between particles results in a repulsive force, namely the dispersive force. The dispersive force keeps bed load particles an average distance apart from each other during their course of motion. Nevertheless, collision of particles results also in a great resistance to the flow. The mechanism of the dispersive force is illustrated, in general, by the following example (Wang & Qian, 1985a). As shown in Figure 2.1, particle P located at the point 1 at instant t_1 moves at a velocity, $\vec{V}$, elative to particle P_1, and it reaches point 2 at instant t_2 after collision with P_1 and its velocity changes to $\vec{V'}$. Such abrupt change in velocity, both in magnitude and direction, because of collision, causes an acceleration. The average acceleration during the time interval $t_2 - t_1$ is

$$\vec{a} = \frac{(\vec{V'} - \vec{V})}{(t_2 - t_1)} = \frac{\Delta \vec{V}}{\Delta t} \tag{2.1}$$

According to Newton's second law, particle P must be subjected to action of a force. An average value of the force can be given by

$$\vec{f} = M\vec{a} = \frac{M((\Delta \vec{V} \cdot \vec{i})\vec{i} + (\Delta \vec{V} \cdot \vec{j})\vec{j})}{\Delta t} \tag{2.2}$$

where M is the mass of the particle P, $\vec{i}$ and $\vec{j}$ are the basic vectors in the longitudinal and vertical directions, respectively. $\Delta \vec{V}$ and its two components are shown in Figure 2.1. They represent the force $\vec{f}$ and two force components in the longitudinal and the vertical directions. The dispersive stress is the sum of such forces acting on the particles in unit area. The component in the flow direction, T, and the vertical direction, P, of the dispersive stress are respectively referred to dispersive shear stress and dispersive pressure.

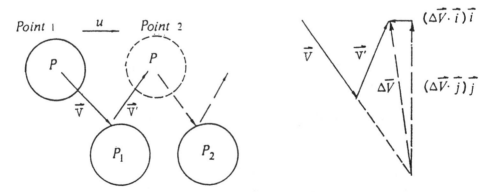

Figure 2.1. The dispersive force as a result of collisions between moving particles.

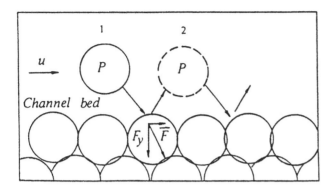

Figure 2.2. Reactionary force on the bed particles.

The dispersive pressure differs from the fluid pressure in its orientation. For cohesionless particles on channel bed, the fluid pressure can be transmitted through fluid in gaps between particles and acts on every part of each particle. This results in nothing but the buoyancy force. Nevertheless, the dispersive pressure acts only on the contact part and has different nature than the fluid pressure. As shown in Figure 2.2, particle P collides with the particles on the channel bed, and is acted by the dispersive force which is in the direction of $\overrightarrow{\Delta V}$. In the meantime the particles on the bed are acted upon by a reactionary force $\overrightarrow{F}$. The vertical component of the force, F_y plays an important role in stability of the channel bed. It is well known that the initiation of sediment particles on a flat bed is mainly related to the lift force of the flow. F_y counterbalances a part of the lift force and thereby prevents the particles from entering into motion. With moving bed load increasing following increasing flow velocity, there comes an instant that the average value of F_y equals the lift force. The channel bed cannot be eroded further and the bed load motion reaches an equilibrium. The bed load layer acts like a protective coating, which protects the bed against continuous erosion. The higher the flow velocity and the larger the tractive shear stress, the more the bed load and the thicker the protective coating, so that the channel bed can maintain stable (not being continuously eroded) at different hydraulic conditions.

Bagnold (1954) abstracted the concept of dispersive force from his experiment on gravity-free dispersion of spheres in a Newtonian fluid under shear. Based on the experimental results the following expressions were obtained:

$$P = a_i \rho_s \cos \alpha \left(\lambda D \frac{du}{dy} \right)^2 \tag{2.3}$$

$$T = a_i \rho_s \sin \alpha \left(\lambda D \frac{du}{dy} \right)^2 = P \tan \alpha \tag{2.4}$$

$$\lambda = \frac{1}{\left(\dfrac{S_{vm}}{S_v}\right)^{\!1\!/\!3} - 1} \tag{2.5}$$

Equations (2.3) and (2.4) are valid only if the inertia of the solid particles dominates the two phase flow, or if the dimensionless number

$$N = \frac{\rho_s D^2 \left(\dfrac{du}{dy}\right) \sqrt{\lambda}}{\mu_0} \tag{2.6}$$

is larger than 450. In the formulas above, a_i is a constant, α frictional angle of moving particles, D and ρ_s diameter and density of particles, du/dy velocity gradient, μ_0 the viscosity of the fluid, S_v volume concentration of particles and S_{vm} the maximum volume concentration when the particles pile together and contact each other, and λ is called linear concentration.

Bagnold (1956) suggested that the submerged weight of bed load is supported by the intergranular dispersive pressure P. The dispersive shear stress T composes an important part of the resistance. Bagnold (1956) and Takahashi (1978) assumed gravels in debris flow move as bed load and are supported by the dispersive force. They developed velocity distribution formulas for debris flow by using the expressions of the dispersive force.

Moreover, if the energy slope is sufficiently high and the bed of a channel is composed of cohesionless, loose grains, the bed material may be sheared and move in layers. No turbulence can develop in the extremely hyperconcentrated flow and is called 'laminated load motion' (Wang & Qian, 1985). Laminated load motion is a high intensity of bed load motion because the submerged weight of particles is supported by the dispersive force and contact force. Resistance to the flow is contributed mainly by the dispersive shear stress and is relatively high.

2.1.2 *Suspended load*

Sediment is referred to suspended load if it is hold in suspension by turbulent velocity components. The submerged weight of suspended load is supported by diffusive force which is associated with turbulence intensity and concentration gradient of sediment. According to Fick's law, if there is a concentration gradient in a turbulent flow field, there must be diffusive movement. The amount of sediment transported through unit area due to diffusion is given by

$$\vec{q} = \varepsilon_s \nabla S_v \tag{2.7}$$

where ε_s is sediment-diffusion coefficient and ∇S_v concentration gradient of suspended sediment. Only in a homogeneous turbulent flow is ε_s a constant. In shear turbulent flow ε_s is assumed to be proportional to the turbulent momentum-diffusion coefficient or eddy viscosity, ε_m, i.e.

$$\varepsilon_s = \beta\varepsilon_m \tag{2.8}$$

where β is a factor describing difference of diffusion of a discrete sediment particle from the diffusion of fluid. The value of β remains uncertain in flows with different sediment. Carstens (1952) discussed the value of β, analyzing measured data for the oscillatory motion of a spherical particle in fluid, and obtained $\beta < 1$ for all experimental cases. On the other hand, some investigators have experimentally indicated $\beta > 1$; for example, Ismail (1952) showed that β in a cylindrical pipe flow varies with a function of sediment size and larger than 1, Singamsetti (1966) found that the analyzed value of β in a sediment turbulent jet falls in the range of 1.2-1.5 and increases with the increase of the boundary Reynolds number, and Wang & Qian (1984) obtained $\beta = 1.3$ for flows with $D = 0.15$ mm fine sand in a closed conduit. Jobson & Sayre (1979) have given experimental evidence that β in an open-channel sediment-laden flow depends on the turbulent characteristics, and suggested possibilities of $\beta < 1$ and $\beta > 1$. Umeyama (1992) analyzed the data measured by Vanoni (1946) and Einstein & Chien (1955) and obtained that β falls in the range 1.05-1.30 for Vanoni's data and in the range 0.4-0.8 for Einstein & Chien's data. Nevertheless, all the investigators accept β as a constant in flow with a selected sediment.

In a two-dimensional open channel flow,

$$\varepsilon_m \frac{du}{dy} = U_*^2 \left(1 - \frac{y}{H}\right) \tag{2.9}$$

By using the traditional logarithmic velocity-defect law, the distribution of sediment-diffusion coefficient is given by

$$\varepsilon_s = \beta\kappa U_* y \left(1 - \frac{y}{H}\right) \tag{2.10}$$

where κ is the von Kármán constant, H is flow depth, U_* is shear velocity.

Equation (2.10) is not valid near the surface of the open channel because it gives zero sediment diffusion coefficient and zero sediment concentration that conflict with the measurements of sediment concentration distribution, especially in hyperconcentrated flows. Modification has been made to coincide with the measured data (Wang & Qian, 1984):

$$\varepsilon_s = \begin{cases} \beta\kappa U_* y(1 - y/H), & y/H \leq a \\ \beta\kappa U_* Ha(1 - a), & y/H > a \end{cases} \tag{2.11}$$

where $a = 0.7$. By substituting the expression into the sediment diffusion equation, a concentration distribution formula is obtained which is still valid in hyperconcentrated flows and pipe flows.

Suspended sediment in natural streams have a wide sieve curve. The larger sizes down to some minimum size are similar to those already composing the bed so that the transport rate can be adjusted to the capacity rate of the flow. The finer fractions, which are more easily kept in suspension are not so likely to be found in the bed, are transported under conditions less than the sediment-carrying capacity of the flow. As a result, the transport of this portion is independent of flow and depends only on conditions extraneous to the flow. This fine material has been given the name wash load to distinguish it from the bed-material load. By rule of thumb, the critical diameter of wash load is d_{b5} or d_{b10}, where d_{b5} or d_{b10} are the diameters of 5% or 10% of sediment on the channel bed is finer. This criterion is convenient but it is not valid in hyperconcentrated flows. During hyperconcentrated flood, sediment much coarser than d_{b10} can be transported over several hundreds of kilometers without deposition, because fall velocity of the sediment is remarkably reduced. It is essentially washed through the channel and should belong to the category of wash load. With a sophisticated model of suspended load motion, Wang & Dittrich (1992) suggested a new criterion, the Rouse number Z, to differentiate wash load and bed material load,

$$Z = \frac{\omega}{\beta \kappa U_*} \tag{2.12}$$

where ω is group fall velocity of sediment. The Rouse number Z denotes the relative intensity of settlement with respect to the diffusion of suspended sediment. It was proposed that sediment carried by a flow is referred to wash load if $Z < 0.06$, to bed material load if $0.1 < Z < 3$, and to bed load if $Z > 5$. Bed material load has significant influence on channel bed deformation and often exchange with bed load and bed sediment. Transition from wash load to bed material load falls in the range $Z = 0.06$-0.1, and transition from bed material load to bed load falls in the range $Z = 3$-5.

2.1.3 *Neutrally buoyant load*

In a hyperconcentrated flow presence of clay and silt endorses the sediment-water mixture a non-Newtonian fluid with yield strength. A part of sediment can be supported by yield stress and does not settle down. Such a part of sediment is called neutrally buoyant load. There is no relative movement between neutrally buoyant load and fluid. If a hyperconcentrated flow carries only neutrally buoyant load, it is a pseudo-one phase flow.

Let τ_B denote yield stress and D_o the maximum diameter of particles that can be supported by yield stress. We have

$$D_o = K \frac{6\tau_B}{\gamma_s - \gamma_f} \tag{2.13}$$

where γ_s is specific weight of sediment, γ_f buoyancy force on unit volume of sediment particle. γ_f equals specific weight of liquid or clay suspension if one sediment particle of diameter much larger than clay particles is involved, and can be taken as specific weight of a mixture of liquid and sediment excluding grains larger than $D_o/50$ if sediment in the mixture has a wide sieve curve (Wang, 1987). For instance, if one considers gravels in suspension of clay and silt, the buoyancy force acting on one gravel equals the volume of the gravel by the specific weight of the suspension. If buoyancy force on gravel in debris flow is concerned, however, particles of diameter smaller than $1/50$ of the size of the gravel should be taken into account. K is a constant and many results indicated $K \approx 1$.

In a non-Newtonian hyperconcentrated flow a part of sediment of diameter smaller than D_o is neutrally buoyant load. In hyperconcentrated flows occurring in the Kuye River, the Huangfuchuan River and other tributaries of the Yellow River with origin in the coarse sediment areas, only a small part of sediment is neutrally buoyant load because the fluid exhibits small yield stress. In hyperconcentrated flows occurring in some tributaries of the Yellow River with origin in the fine sediment areas, such as the Weihe River, and mud flows and some debris flows consisting mainly of clay, sand and gravels, yield stress of the fluids is quite high and D_o is large, consequently much more sediment, e.g. more than 50%, are transported as neutrally buoyant load.

2.2 PATTERNS OF MOTION OF HYPERCONCENTRATED FLOW

2.2.1 *Patterns of motion*

(a) *Suspended load + bed load motion.* In hyperconcentrated flows in Northwest China, sediment moves mainly as suspended load along with a small part of bed load. Suspended load motion consumes only turbulent energy and does not cause additional resistance to the flow but bed load does. Hyperconcentrated flow occurs when heavy rainstorm causes erosion of the Loess Plateau and brings much fine sediment into rivers. The rate of sediment transport can not be estimated from hydraulic parameters because quite a lot of the sediment is wash load.

(b) *Suspended load + neutrally buoyant load motion.* If hyperconcentrated flow exhibits yield stress a part of fine sediment moves as neutrally buoyant load, while coarse sediment is transported as suspended load and bed load. Some hyperconcentrated flows occurring in the Weihe River, some subviscous debris flows composed of fine sediment in Southwest China and hyperconcentrated flows in the North Fork Toutle River (Pierson & Scott, 1985) are of such pattern of motion.

(c) *Neutrally buoyant load motion.* If a flow carries enough clay material, the mixture may exhibits strong yield strength and most sediment in the flow belongs to neutrally buoyant load. Mud flow in the Loess Plateau is an example of such pattern. Because viscosity of the mixture is quite high, a relatively high energy slope is needed to maintain the flow.

(d) *Laminated load motion and neutrally buoyant load + laminated load motion*. If energy slope of a flow is sufficiently high and there is only cohesionless material available, laminated load motion may develop. Water debris flow is essentially a laminated load motion. On the other hand, in viscous debris flow gravels, cobbles and big stones may move as laminated load and sand and silt may be neutrally buoyant load.

2.2.2 *Major factors controlling patterns of motion*

(a) *Sources of sediment*. Hyperconcentrated flow transports a huge amount of sediment. There must be an abundant source of sediment in hyperconcentrated flow area. For instance, the Loess Plateau consists of 10-200 m deep loess deposit with loamy texture and poor vegetative cover. It is liable to be eroded during rainstorm and provides rivers with huge amount of sediment and consequently results in hyperconcentrated flows very often. Eruptions of the Mount St. Helens, Washington produced much volcanic debris on the hillslope. On the evening of March 19, 1982 an explosive eruption melted snow and ice, the water flowed out of the crater, initiated lahar (volcanic debris flow), fed a lot of sediment to the North Fork Toutle River and resulted in hyperconcentrated flows (Pierson & Scott, 1985). In the northern alpine area, Quaternary debris deposits, with sizes ranging from clay to gravels, covered the hillslope. Massmovement often took place during rainstorm in forms of landslide and earth flow (solid materials, saturated with water, moved down the slope as a series of layers relatively slipping, that can be referred as a kind of laminated load motion), that provided sediment sources for debris flow and hyperconcentrated flow (de Jong, 1992).

(b) *Sediment constituents*. The patterns of hyperconcentrated flow depend, to a great extent, on rheological properties of water-sediment mixture. Rheologic behavior of the mixture depends in turn on size distribution and mineral compositions of sediment. Mineral compositions play a more important role in rheology of sediment suspension than size distribution (see Chapter 4). For example, sediment suspension exhibits yield strength only as it consists of clay material. Nevertheless, content of clay in natural sediment has some correlation with size distribution, because clay particles are generally much finer than non-clay particles. Fei (1983) found that the minimum concentration for sediment suspension to exhibit yield stress, S_{vo}, is closely related to the content of sediment finer than 0.01 mm, as shown in Figure 2.3. If sediment consists of 70% fine particles a suspension exhibits yield stress as long as the concentration is higher than 4%,

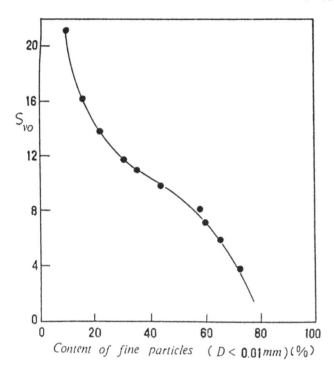

Figure 2.3. The minimum concentration for the suspension to exhibit yield stress as a function of the content of fine sediment.

whereas if the content of fine particles is lower than 10% a suspension remains Newtonian with volume concentration up to 20%. The relation is not universal but similar relations can be expected for sediments from selected areas. For instance, clay particles in the Yellow River basin are composed mainly of illite, with small regional differences. A good relationship between content of fine sediment and rheologic behavior of suspension may exist.

(c) *Hydraulic factors.* High velocity, turbulence intensity and energy slope are needed for initiating and maintaining hyperconcentrated flow. In Northwest China hyperconcentrated flow occurs only in flood season because initiation of so much sediment needs high intensity of turbulence. Following decrease in velocity and turbulence intensity during descending limb of a flood, much sediment deposits and brings about serious channel deformation sometimes. Neutrally buoyant load motion and laminated load motion need high energy slope because such patterns of hyperconcentrated flow involves high resistance. A study on initiation of debris flow indicated that γqJ must be larger than a critical value so that a debris flow can be initiated if there are enough debris materials on the channel bed, in which q is the flow discharge per width and J is the energy slope (Wang & Zhang, 1990).

2.2.3   *Patterns of motion of hyperconcentrated flow in the Yellow River basin*

There is a distinct regional difference in grain size distributions in the middle reaches of the Yellow River. Towards the south-east the median diameter of sediment decreases from 0.045 mm to 0.015 mm. According to the size distributions, the middle reaches of the Yellow River can be divided into two regions, namely, the regions of coarse sediment and fine sediment. Hyperconcentrated flow in the coarse sediment region are mostly two phase flow and that in the fine sediment region are sometimes pseudo-one-phase flow. Table 2.1 lists hyperconcentrated flows and patterns of motion in different rivers in the area (Qian & Wan, 1986). Sediment from the Kuye River was coarse and the content of particles finer than 0.01 mm was only 1-3%. Most of sediment in hyperconcentrated flow was suspended load and bed load. However, the role played by the fine particles in reducing fall velocity of coarse sediment and maintaining hyperconcentrated flow should not be ignored. In the Luohui Irrigation Canal, however, the flow is sometimes pseudo-one phase because the content of fine sediment was rather high. The sediment suspension exhibited high yield stress and most of the sediment became neutrally buoyant load. Hyperconcentrated flows in the Yellow River is two-phase flow and the content of fine particles is in the range 1-25%. Neutrally buoyant load consists only a small part of the sediment, and suspended load consists the main part of the solid phase. It should be pointed that the yield stress was estimated from measured value with viscometers, but the shear rate in the viscometers were much higher than the shear rate in hyperconcentrated flow in nature. Most sediment suspensions are somewhat pseudoplastic, in other words, the suspensions exhibit smaller yield stress and higher rigidity coefficient at low shear rate and larger yield stress and lower rigidity coefficient at high shear rate. Therefore, the estimation of percentage of neutrally buoyant load in the table is on the higher side.

Table 2.1. Patterns of motion of hyperconcentrated flows in the Yellow River basin

River	Gauging station	Concentration (kg/m^3)	Content of fine sediment (%)	D_o (mm)	Percentage of neutrally buoyant load (%)	Patterns of motion
Kuye River	Shenmu station	1200	1-3	0.2-0.24	23-30	Two phase flow, suspended load + bed load
		800	3	0.02	5	Two phase flow suspended load + bed load
Luohui irrigation canal	Yijing east main canal	800	10-47	0.1-0.33	95-100	Pseudo-one-phase flow neutrally buoyant load
Weihe River	Lintong station	650-800	16-20	0.04-0.11	55-92	Two phase, sometimes pseudo-one-phase flow

REFERENCES

Bagnold, R.A. 1954. Experiments in a gravity-free dispersion of large spheres in a Newtonian fluid under shear. *Proc. Royal Soc. London, Ser.A*, Vol. 225: 49-62.

Bagnold, R.A. 1956. The flow of cohesionless grains in fluids. *Phil. Trans. Roy. Soc. London*, Vol. 249.

Carstens, M.R. 1952. Accelerated motion of spherical particles. *Trans., AGU*, Vol.33(5): 713-721.

de Jong, C. 1992. A catastrophic flood/multiple debris flow in a confined mountain stream: an example from the Schmiedlaine, southern Germany. *Proc. of Chengdu Symp. on Erosion, Debris Flow and Environment in Mountain Regions*, IAHS Publ. no. 209: 237-245.

Einstein, H.A. & N. Chien 1955. *Effects of heavy sediment concentration near the bed on the velocity and sediment distribution*. Report No.8, Univ. of Calif. Berkeley, Calif.

Fei, X. 1983. Grain composition and flow properties of heavily concentrated suspensions. *Proc. 2nd Intern. Symp. on River Sedimentation*, pp. 307-308.

Ismail, H.M. 1952. Turbulent transfer mechanism of suspended sediment in closed channels. *Trans., ASCE*, Vol. 117: 409-446.

Jobson, H.E. & W.W. Sayre 1979. Prediction concentration in open channels. *J. of Hydr. Div., ASCE*, Vol.96(10): 1983-1996.

Pierson, T.C. & M. Scott 1985. Downstream dilution of a lahar: Transition from debris flow to hyperconcentrated streamflow. *Water Resources Research*, Vol. 21(10): 1511-1524.

Qian, N. & Z. Wan 1983. *Mechanics of Sediment Transport*. Academic Press (in Chinese).

Qian, N. & Z. Wan 1986. *A critical review of the research on the hyperconcentrated flow in China*. IRTCES, Beijing.

Singamsetti, S.R. 1966. Diffusion of sediment in a submerged jet. *J. of Hydr. Div., ASCE*, Vol. 92(2).

Takahashi, T. 1978. Mechanical characteristics of debris flow. *J. of Hydr. Div., ASCE*, Vol. 104(8): 1153-1169.

Umeyama, M. 1992. Vertical distribution of suspended sediment in uniform open channel flow. *J. of Hydr. Engineering, ASCE*, Vol. 118(6): 936-941.

Vanoni, V.A. 1946. Transportation of suspended sediment by water. *Trans. ASCE*, Vol. 111: 67-133.

Wang, Z. & A. Dittrich 1992. A study on problems in suspended sediment transportation. *Proc. of 2nd Intern. Conf. on Hydraulics and Environmental Modelling of Coastal, Estuarine and River Waters*, Vol. 2: 467-478.

Wang, Z. 1987. Buoyancy force in solid-liquid mixtures. *Proc. of 22nd Cong. IAHR, Lausanne*, Fluvial Hydraulics, 86-91.

Wang, Z. & N. Qian 1984. Experimental study of two phase turbulent flow with hyperconcentration of coarse particles. *Scientia Sinica, Ser. A*, Vol. 27(12): 1317-1327.

Wang, Z. & N. Qian 1985. Experimental study of motion of laminated load. *Scientia Sinica, Ser. A*, Vol.28(1): 102-112.

Wang, Z. & X. Zhang 1990. Initiation and laws of motion of debris flow. *Proc. of Intern. Symp. on Hydraulics/Hydrology of Arid Land, ASCE, San Diego*, 596-601.

CHAPTER 3

The origin and formation of hyperconcentrated flow in the Yellow River basin

3.1 THE GEOMORPHIC, GEOLOGICAL, AND HYDROLOGICAL FEATURES OF THE LOESS PLATEAU REGION OF THE YELLOW RIVER BASIN

The Yellow River flows through a vast loess plateau, covering an area of 236 000 km^2 in its upper and middle reaches. Ever since the plateau was first formed, it has been in the rising stage of tectonic movement. Except rocky mountains with or without forests, deserts, and sandy grasslands, most of the area is covered by loess or red loess with a thickness of dozens meters to 200 m.

Loess is a kind of silt loam, mainly (60-70%) composed of silt of 0.05-0.002 mm in diameter. It contains only 10 to 20% clay particles, so it has a loose structure. In most farmland, the granular structure content is less than 5 or 10%. Granules are connected by calcium carbonate ($CaCO_3$), which comprises about 10-15% of the entire soil. Loess has obvious vertical joints filled with calcium carbonate, which makes loess particles stand vertically and is easily dissolved in water. Loess also has high porosity, generally, 40 to 50%. So after a rainstorm, water easily permeates along the joints and dissolves calcium carbonate, and consequently granules disintegrate and loess collapses. In addition, the permeability of the underlying red loess or rock stratum is less than that of the recent deposits of loess in the top layer. Therefore, ground water usually accumulates along the interface and forms a weak plate, along which lump slide easily occurs.

This area is in an arid/semi-arid zone. The long-term average annual precipitation, varying between 200-700 mm, decreases from southeast to northwest. The distribution of rainfall is rather uneven, both within a year and between years. On the average, 70% of the annual precipitation is concentrated in the period from June to September. Rainstorms occur frequently in summer, and their intensity may reach 1-3 mm/min. Most soil erosion occurs during rainstorms, and quite often flash floods at hyperconcentration result from it.

Due to low, uneven precipitation and over-cultivation, the vegetation in this region is very poor. The primary natural plant cover has already been destroyed.

Figure 3.1. Loess plateau (after Yin Hexian).

Only 3% of the entire area (1.2×10^6 hectares) is covered by natural forest. The coverage of natural grassland is also poor, and it cannot protect soil. Poor vegetative cover, in turn, allows natural calamities such as drought, frost injury, hail, etc. to harm the existing vegetative cover. This vicious circle lasts for a long time and accelerates soil erosion.

Due to these factors, soil erosion in this region is severe, and the whole loess plateau is cut by thousands upon thousands of deep criss-crossed gullies and is broken up into numerous small watersheds of sizes ranging from 10 to 100 km^2. Gullies are as deep as dozens meters to three hundred meters. The large height difference favours gravitational erosion during the flood season.

On the average, there are 3 to 7 kilometers of gullies per square kilometer. The area of the gullies comprises about 40% to the entire area, and about 60 to 80% of the total sediment load is eroded from the gullies. Figure 3.1 is a bird's-eye view of the gullied-hilly loess region.

Heavy loss of soil and water has made this area the source of hyperconcentrated floods. Following rainstorms in summer, flash floods at hyperconcentration occur frequently along tributaries. Sediment from tributaries in this region, with a total area of 266 000 km^2, contributes more than 90% of the total sediment load of the entire Yellow River.

3.2　THE DISTRIBUTION OF THE MODULUS OF SEDIMENT YIELD IN THE LOESS PLATEAU

Most sediment in the Yellow River comes from the loess plateau, of which area with very large modulus of sediment yield is restricted to a comparatively small portion.

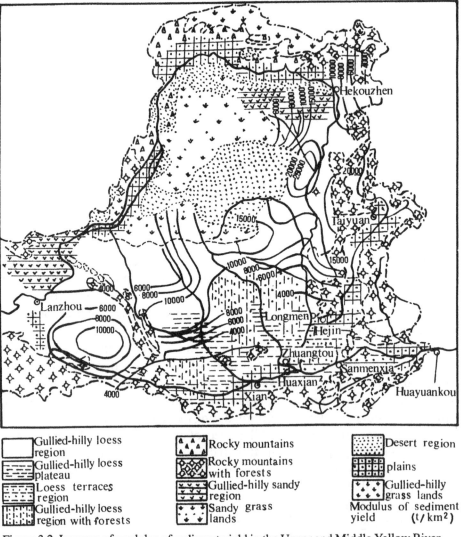

Figure 3.2. Isogram of modulus of sediment yield in the Upper and Middle Yellow River.

Figure 3.2 shows isogram of the modulus of sediment yield in the upper and middle Yellow River (Gong & Xiong, 1987).

The maximum modulus of sediment yield is 25 000 t/km². A principal sediment yield zone is the Beiyu mountain area, from which the Wuding River, the Yanhe River, the North Luohe River, and the Jinhe River originate. The modulus of sediment yield in this area is about 10 000 t/km². The common tendency is that the modulus of sediment yield is large in the north part of the region and decreases gradually toward the south.

3.3 THE VARIATION OF LOESS SIZE AND MAXIMUM CONCENTRATION

In the loess plateau the size distribution of recent loess deposits obviously zonal. The median particle diameter varies from 0.045 mm in the northwest to less than 0.015 mm in the southeast (see Figure 3.3, Qian et al., 1987). Actually, Figure 3.3 is the enlarged lower part of Figure 3.2. In the valleys of the Huangpuchuan and Kuye Rivers and the source zones of the Wuding, North Luohe, and Jinhe Rivers, the median particle diameter of recent loess deposits is 0.045 mm, whereas in the upper reaches of Weihe river it is about 0.015 mm.

Figure 3.4 displays granulometric curves of representative samples taken from these rivers.

In Table 3.1, maximum recorded concentrations taken from different tributaries are listed.

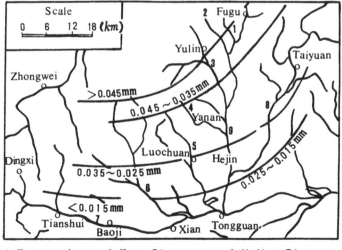

Figure 3.3. Variation of median diameter of new loess in the middle reaches of the Yellow River.

1 Huangpuchuan	2 Kuye River	3 Wuding River
4 Yanhe River	5 North Luo River	6 Jinghe River
7 Weihe River	8 Fenhe River	9 Yellow River

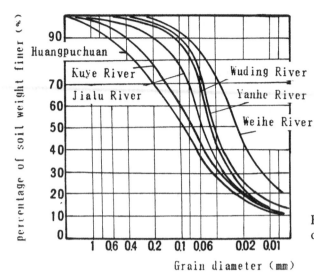

Figure 3.4. Granulometric curves of rivers in the loess plateau.

Table 3.1. Maximum recorded sediment concentration in different tributaries.

River	Gauging station	Annual load 10^8 (t)	Maximum recorded concentration (kg/m^3)
Huangpuchuan	Huangpu	0.641	1570
Gushangchuan	Gaoshi	0.278	1300
Kuye	Yunjiachuan	1.35	1700
Wuding	Beijiachuan	1.06	1290
Qingjian	Yanchuan	0.453	1150
Yanhe	Ganguyie	0.546	1200
Fenghe	Hezing	0.438	386
Weihe	Hianyang	1.68	729
Jinghe	Zhangjiashang	2.86	1430
North Luo	Zhuangte	0.968	1340

Comparing Table 3.1 with Figure 3.2 and Figure 3.3, one may obtain the following concept. In accordance with the variation of the size distribution of loess, the sediment concentration of rivers originating in different regions is different. In general, rivers originating in the northwest, such as Huangpuchuan, Kuye, Wuding, North Luohe, Jinhe, etc. usually are at higher concentrations, and those from the southeast, such as Weihe River, are at lower concentrations. Comparison of Table 3.1 with Figure 3.2 and Figure 3.3 readily validates this generalization.

3.4 LIMITING CONCENTRATION OF HYPERCONCENTRATED FLOW AT DIFFERENT LOCALITIES IN THE BASIN

This problem was studied by analyzing field data accumulated by the runoff experiment station at Chaba Ravine, a small tributary of the Dali River, in Zizhou County, Shannxi Province (Qian et al., 1985). The Chaba Ravine has a drainage area of 187 km^2. The formation of hyperconcentrated flow and the distinguishing features of the flow in the process of confluence were analysed.

Because of the extremely low erosion-resistance capacity of loess, hyperconcentrated flow can easily be formed by the combined action of splash, rill and gravitational erosion. The limiting concentrations, i.e. the possible maximum concentration, of the hyperconcentrated flow at different localities in the basin are listed in Table 3.2. Because the measured maximum concentration may not represent the actual maximum concentration, not the maximum measured concentration was taken as the limiting concentration. Based on all the field data from 1963, 1964, and 1966, the maximum concentrations for each period of runoff were extracted, then the frequency of these maximum concentrations was plotted. The concentration corresponding to 10% frequency is referred to as the 'limiting concentration'. Table 3.2 shows that through splash erosion alone, a concentration of 510 to 690 kg/m^3 was already reached. While overland flow carved out rills and furrows the limiting concentration increased to 860 kg/m^3. An additional supply of sediment from gravitational erosion in the gullied bank area caused the concentration to rise to 990 kg/m^3. The limiting concentrations then increased even further to 1160 kg/m^3 in the tributary and 1290 kg/m^3 in the main stem of the Dali River.

Figure 3.5 presents the hydrographs of sediment concentration in the Chaba Gully Basin on Aug. 15, 1966. Due to long-term severe erosion, a gully system

Table 3.2. Limiting sediment concentration of the flow at different localities in the Dali River basin.

	Locality	Representative station	Limiting concentration (kg/m^3)
Slope of 'Mao'	Small plot without drainage channels	Plot No. 11 of Tuanshan Gully	510
	Upper part of the slope without drainage channels	Plot No. 4 of Tuanshan Gully	690
	Middle part of the slope with rills and furrows	Plots No. 2 & 3 of Tuanshan Gully	860
	The whole slope region including the slope of 'Mao' and the gully bank (trenches and gravitational erosion)	Plots No. 7 & 8 of Tuanshan Gully	990
Gully and river	Branch gully	Tuanshan Gully	920
	Gully	Shejie Gully	920
	Tributary	Chaba Ravine, Dali River	1160
	Main stem of Dali River	Suide	1290

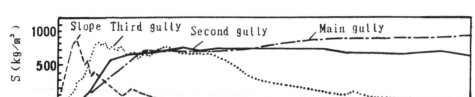

Figure 3.5. Hydrographs of sediment concentration in the Chaba Gully watershed on Aug. 15, 1966.

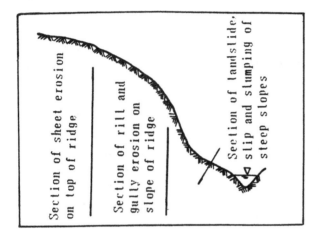

Figure 3.6. Sketch of soil erosion for slopes in the rolling loess region on the middle reaches of the Yellow River.

has fully developed in the watershed. The channels have short lengths and steep slopes. The master gullies are generally 5 to 40 km long, with bed gradient 0.01-0.03 mostly cutting into the bedrock. Rills and second-order gullies have even steeper gradients reaching 20% or more. Sideward erosion, undercutting of gully beds, and retrogressive erosion of gullies usually occur during rainstorms. Under natural conditions, all the gullies here are sediment-transporting channels, with little possibility of deposition. On the contrary, more sediment is added as a result of the gravitational erosion of the gully bank. Based on synchronous observations at Zizhou Experimental Runoff Station in 1964 and 1967, the modulus of soil erosion at the top of loess mounds and ridges (i.e. near the divide) is 247 t/km^2, whereas that on the slope of ridges is 13800 t/km^2 and that in the gully itself becomes as large as 21 100 t/km^2.

The mechanism of sediment yield from slopes to gullies was described by Gong & Xiong (1987) as follows:

1. Loose-structured loess becomes disintegrated and scattered after being splashed and soaked by rain water.

2. Together with the formation of surface runoff the soaked and disintegrated

loess gradually gets mixed with the water and is carried away by the runoff (the soil is first soaked and then runoff begins).

3. The sediment-laden flow grows gradually in the process of moving along the slope surface. Trenching and undercutting, streams then coalesce to form larger channels. Funnel-like holes initiated by soil erosion appear.

4. As a result of the recent development of gullies and drainage networks the land surface gradient is gentle at the top and steep at the foot of the slope (see Figure 3.6). Hence, the soil eroded away from the top of the slope has no chance to deposite on the downstream slope. On the contrary, gravitational erosion in the form of slumping and slipping at the foot of steep slopes and cliffs adds sediment to the flow.

Thus it can be assumed that all the sediments washed away from the slopes and gullies in the crisscrossed loess regions are delivered into the rivers via systems of gullies of different orders. In other words, the sediment delivery ratio of small-gullied watersheds is close to 100%, irrespective of the size of the catchments. This is quite different from the situation in other regions.

3.5 DISCHARGE-SEDIMENT TRANSPORT CAPACITY RELATIONSHIP FOR GULLIES OF DIFFERENT ORDERS

In the upper and the middle part of the slope, the drainage area at the top is so small that the hydrographs of surface runoff and sediment concentration are essentially in phase. With the enlargement of catchment area, flow from different parts of the watershed reaches the outlet of the basin at different times. In the recession limb the flow at the outlet of the basin is well advanced to the surface flow from distant parts of the watershed, formed earlier under more intensive precipitatiion and therefore with higher concentration. Therefore the hydrograph of sediment concentration begins to lag behind the hydrograph of surface runoff, and relatively high sediment concentration may be sustained for a much longer time than the peak discharge. For this and other reasons, hardly any correlation between discharge and the sediment concentration at low flow can be found.

Nevertheless, when heavy rainfall pours down over the entire basin and results in torrents of large discharge, flow everywhere is saturated with sediment. The concentration of the flow can not be far from the limiting concentration of that locality. We have already shown in Table 3.2 that from the slope to the various size gullies, the range of variation of the limiting concentration is rather small. Consequently, the concentration of the outflow of a basin at high discharges also remains within a narrow range.

The specific relationship between discharge and concentration causes the functional correlationship between discharge and sediment transport capacity of the loess-hilly areas to assume a form of its own, as shown in Figure 3.7 (Qian et al., 1985). The discharge (Q) and sediment transport rate (Q_s) relationships for the

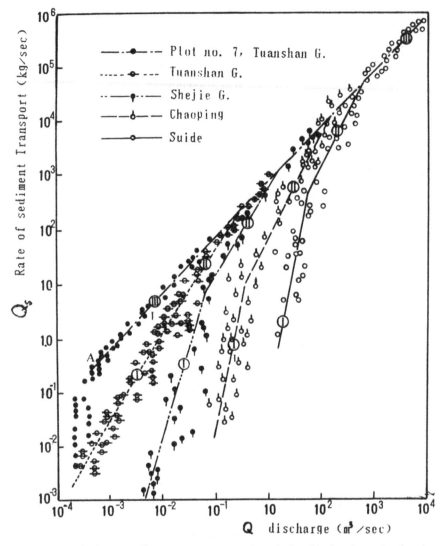

Figure 3.7. Discharge-sediment transport capacity relationship for channels of various orders in the Dali River basin in 1964.

whole slope region (represented by Plot. No. 7 of Tuanshan Gully) and for channels of different orders are plotted in the same diagram. The measured points from each source area collect together in a group. Each group can be subdivided into three lines, as marked in the figure. Along line I all the concentrations are smaller than 400 kg/m³. Along line II all the sediment concentration are larger than 400 kg/m³. In zones I and II, the concentrations at low and intermediate discharges vary to a great extent, and the higher the order of the drainage channel is, the larger the exponent m in the expression $Q_s - Q^m$ will be. Entering zone III, data for various channel orders all merge together into a single line A-A, with a slope of 1:1.

The preceding relationship is completely different from that of ordinary sediment-laden flow, which generally has an *m* value of 2.

3.6 SIZE COMPOSITION FEATURES OF THE SUSPENDED SEDIMENT

In hyperconcentrated flow the size composition of the suspended sediment exhibits some distinct features:

1. As mentioned above, in tributaries originating in the northwest, where the new deposits of loess are coarser, the maximum recorded concentration is larger, while in tributaries originating in the southeast, where the new deposits of loess are finer, the maximum recorded concentration is smaller.

2. At a certain gauging station, higher the concentration, coarser the suspended sediment. Figure 3.8 shows examples taken from the Weihe River and Wuding River. Actually, it is a common trend for the whole region.

3. When the total concentration exceeds a certain value, clay content no longer increases with concentration but rather maintains a certain value. Figure 3.9 and 3.10 are examples taken from Chaba Gully and Huangpuchuan, respectively. The same is true for the Wuding River and other rivers.

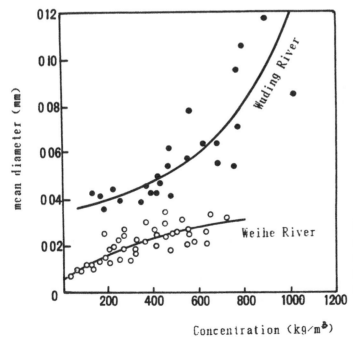

Figure 3.8. The relationship between median diameter and concentration of suspended sediment.

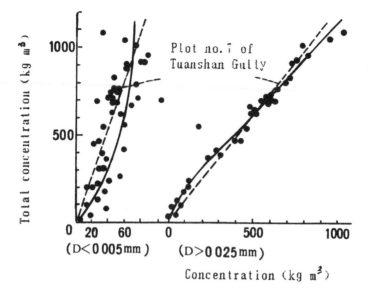

Figure 3.9. The amount of material of various sizes vs. the concentration of suspended sediment for Chaba Ravine.

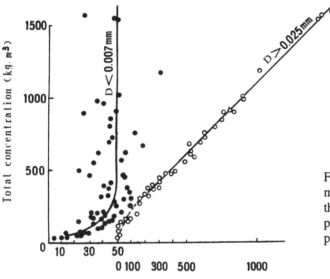

Figure 3.10. The amount of material of various sizes vs. the concentration of suspended sediment at Huangpuchuan.

The persistence of fine-material concentration is a general characteristic not only of the hyperconcentrated flow in river system, but also of the hyperconcentrated lahar-runoff flow. Figure 3.11 shows the sediment concentration of the 1982 lahar-runoff flow in the Toutle River at Tower Road, Mount St. Helens, Washington (Scott & Dinehart, 1985). It can be seen that the concentration of silt and clay (< 0.062 mm) remained in the range of 250-350 kg/m³ for over 3 hours.

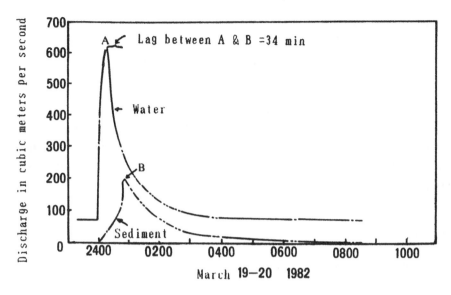

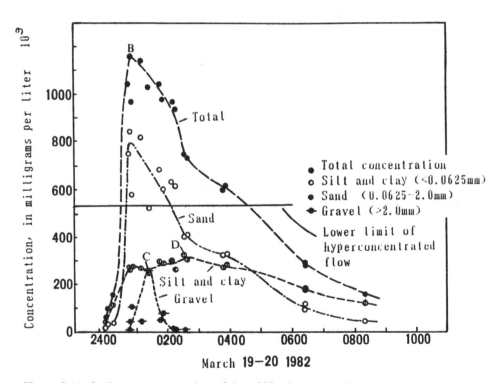

Figure 3.11. Sediment concentration of the 1982 lahar-runout flow in the Toutle River at Tower Road.

The features noted above are important in the transport of hyperconcentrated flow. A certain amount of fine particles form an intricate network of floc structure which effectively reduces the settling velocity of coarse particles, thereby ensuring a high sediment transport capacity. When the concentration rises beyond a certain limit further increases in concentration will only make the sediment composition coarser; and the clay content will not increase further. As will be discussed later, yield stress of the hyperconcentrated fluid increases with the high power of the clay content. The clay content does not increase in proportion to total concentration, thereby ensuring that the flow will not transform into a laminar one which requires a much larger slope to be kept in motion.

REFERENCES

Gong, S. & T. Jiang 1978. Water-soil loss from small watersheds on the loess plateau and its protection. *Scientia Sinica*, 1978.

Gong, S. & G. Xiong 1987. The origin and transport of sediment of the Yellow River. *Selected Papers of Researches on the Yellow River and Present Practice, Yellow River Conservancy Commission*, pp. 28-38.

Qian, N. (Ning Chien), K. Wang, l. Yan & R. Fu 1987. Source of coarse sediment in the Middle Yellow River Basin and its effect on aggradation and degradation downstream. *Selected Papers of Researches on the Yellow River and Present Practice, Yellow River Commission*, pp. 39-46.

Qian, N. (Ning Chien), R. Zhang, Z. Wan & X. Wang 1985. The hyperconcentrated flow in the main stem and tributaries of the Yellow River. *Proc. of the International Workshop on Flow at Hyperconcentrations of Sediment, Publication of International Training and Research Centre on Erosion and Sedimentation*, pp. 3-4-20.

Scott, K.M. & R.L. Dinehart 1985. Sediment transport and deposit characteristics of hyperconcentrated streamflow evolved from lahars at Mount St. Helens. *Proc. of the International Workshop on Flow at Hyperconcentrations of Sediment, Publication of International Training and Research Centre on Erosion and Sedimentation*, pp. 3-2-33.

CHAPTER 4

Rheological properties of hyperconcentrated flow

4.1 RHEOGRAM

The relation between the shear stress τ and the shear rate du/dy of a moving fluid is a good representation of rheological properties of the fluid. The formula used for describing the relationship is called the rheologic equation and the diagrammatic expression of the formula is called the rheogram. The rheograms of some typical time-independent fluids (or fluids without memorial effect) are plotted in Figure 4.1. For a Newtonian fluid (curve A) τ is proportional to shear rate du/dy:

$$\tau = \mu \, du/dy \tag{4.1}$$

in which μ is the viscosity of the fluid. For a Bingham fluid,

$$\tau = \tau_B + \eta \, du/dy \tag{4.2}$$

in which η and τ_B are called respectively the coefficient of rigidity (or plastic viscosity) and the Bingham yield stress. It corresponds to curve B (in Figure 4.1), a stright line which does not pass through the origin. Equation (4.2) is purely an empirical formula for fitting the experimental data. Bingham fluid remains stationary under the action of a shear less than τ_B. It starts moving when the shear surpasses τ_B.

Curves C and D correspond to pseudo-plastic and dilatant fluid respectively. Their rheologic formula is:

$$\tau = K \, (du/dy)^m \tag{4.3}$$

in which K is the coefficient of consistency and m, the plastic index. For pseudo-plastic fluid, $m < 1$; and for dilatant fluid, $m > 1$.

In some cases rheologic equations with three or even more parameters are used to describe the rheological property of a moving fluid. One of the examples is pseudo-plastic with yield stress. Its rheologic equation, which corresponds to curve E in Figure 4.1, is as follows:

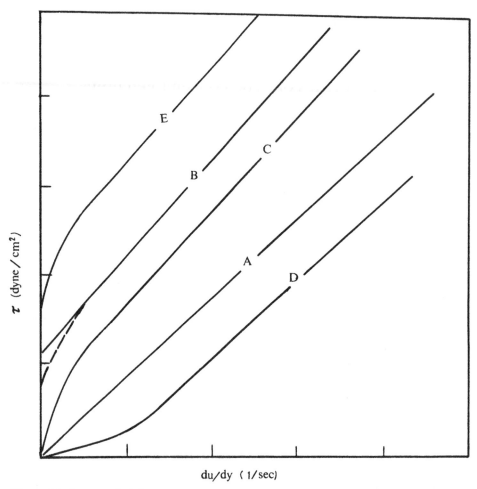

Figure 4.1. Some typical rheograms.

$$\tau = \tau_B + K\,(\mathrm{d}u/\mathrm{d}y)^m \tag{4.4}$$

As a general law, the more the rheological parameters are used, the rheological equation can fit the experimental data more properly, but the equation is more difficult to be used.

Some fluid behaves as time-dependent one, that is, for such fluid there is no unique correlationship between the shear stress and the shear rate, the correlationship varies and depends on the history of the fluid exerted by force. In other words, for such fluid there is some memory effect. At a certain shear rate $\mathrm{d}u/\mathrm{d}y$, shear stress for a thixotropic fluid deteriorates with increasing time, shear stress for a antithixotropic fluid increases with increasing time.

4.2 RHEOLOGICAL MODELS OF HYPERCONCENTRATED FLOW

It is well known that clear water is a Newtonian fluid with a viscosity of μ. Water with low concentration of sediment remains Newtonian fluid, but the viscosity increases with increasing concentration.

At very low concentration, distance between particles is rather large. The existence of each particle brings a certain effect on the motion of surrounding fluid. The effect weakens with distance. Due to the large distance the effect of a particle on its neighbouring particles can be neglected. To any point in the fluid the effect of all the solid particles can be considered as the sum of the effects exerted on it by all the individual particles. Starting from this point of view, Einstein deduced the following famous formula:

$$\mu_r = 1 + 2.5\,S_v \tag{4.5}$$

in which μ_r represents the ratio of the viscosity of turbid water to the viscosity of clear water under the same temperature and S_v is the concentration in volume.

With increasing sediment concentration, the existence of each particle will influence its neighbouring particles. In such case the foregoing formula should be revised. Usually it is written as a multinomial of concentration S_v:

$$\mu_r = 1 + k_1 S_v + k_2 S_v^2 + k_3 S_v^3 + \ldots\ldots$$

in many cases the term of S_v^2 is preserved and terms with higher orders of S_v can be neglected.

$$\mu_r = 1 + k_1 S_v + k_2 S_v^2$$

In most of the studies k_1 remains 2.5, but k_2 values were quite different in different formulas. For instance, k_2 is 2.5 in formula proposed by H. De Bruijin, J.M. Burgers and N. Saite, but 14.4 in formula proposed by E. Guth, R. Simha and O. Gold.

As sediment concentration exceeds a certain value, particularly for sediment containing clay particles, the water- sediment mixture behaves no longer as a Newtonian fluid. The critical concentration varies with sediment constituent, mineral composition of the sediment as well as the water quality. It will be discussed later.

Data of rheological measurements indicated that most hyperconcentrated flows from the middle reaches of the Yellow River can be described as a Bingham fluid (Wan et al., 1979; Zhang et al., 1980; Wang et al., 1983). Matrix of debris flow, that is, the mixture of water and finer granular particles carried by the debris flow, can also be described as a Bingham fluid (Sheng & Xie, 1983; Wang, 1982). In western countries Bingham fluid is also a popularly adopted model for the rheological characteristics of clay slurries (Bird et al., 1982). Although in a few cases pseudo-plastic model was used to describe the rheological characteristics of

Table 4.1. Rheological models of slurry.

Material	Rheological equation	Author(s)	Year
Loess deposits/water	Bingham	Y. Sha	1947
Deposits in reservoir at estuary, in river/water	Bingham	N. Qian & H. Ma	1958
Deposits in river/water	Pseudo-plastic	Y. Zhou	1963
Loess deposits/water	Bingham	Z. Wan et al.	1979
Loess deposits/water	Bingham	H. Zhang et al.	1980
Clay/water	Bingham Pseudo-plastic	J. Dai et al.	1980
Loess deposits/water	Bingham	M. Wang et al.	1983
Bentonite/water, kaoline/water	Bingham	Z. Wan	1982
Loess deposits/water	Bingham	X. Fei	1982, 1983
Slurry of debris flow	Bingham	Y. Wang	1982
Slurry of debris flow	Bingham	S. Sheng & S. Xie	1983
Deposits in bay, reservoir, kaoline, limestone	Bingham	C. Migniot	1968
Clay/water	Bingham	H. E. Babbit & D. H. Caldwell	1940
Kaoline/water	Bingham	D. G. Thomas	1961, 1963
Clay/water	Bingham	G. W. Govier & M. D. Winning	1948
Clay/water	Bingham	N. I. Heywood & J. F. Richardson	1978
Clay/water	Bingham, Crowley-Kitzes	D. G. Thomas	1962
Clay/water	Bingham	B. A. Firth & R. J. Hunter	1976
Clay/water	Bingham	K. Emeya	1970
Clay/water	Bingham	H. D. Weymann & M. C. Chuang, R. A. Ross	1973
Clay/water	Bingham	H. D. Weymann	1965
Clay/water	Bingham	G. F. Brooks & R. L. Whitmore	1968
Clay/water	Bingham	V. M. Dobrychenko et al.	1975
Clay/water	Bingham	E. C. Bingham & T. C. Durham	1911
Clay/water	Bingham	H. Pazwash & J. M. Robertson	1975

clay slurry (Dai et al., 1980). Part of these results of rheological measurement are summarized in Table 4.1.

It can be seen from the table that Bingham model is the most popular one used for describing the rheological property of slurry. Bingham model is simple in form and easy to be treated in mathematical deduction. It should be pointed out that the rheogram of a Bingham model and that of a pseudo-plastic model with properly chosen rheological parameters coincide with each other except that part in the region of little shear rate. Consequently, with the exception of problems involving rheological properties in lower shear rate region both models can be used properly in most cases (Wasp, 1977).

Yield stress τ_B and rigidity η of a Bingham fluid increase with sediment concentration increasing. Researchers set up various empirical formulas for describing the correlationship between rheological parameters and sediment concentration. Exponential formulas are widely used ones, such as:

$$\tau_B = KS_v^m \tag{4.7}$$

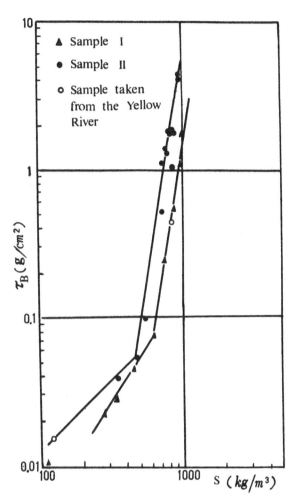

Figure 4.2. $\tau_B \sim S$.

In many cases $m = 3$ (Thomas, 1961; Wan, 1982) is adopted, but in some cases m is smaller in low concentration region but larger in high concentration region, see Figure 4.2 (Wan et al., 1979).

It is found that K varies with mineral content and sediment constituent in a wide range. So it is important to find out the law governing variation of the rheological parmeters.

4.3 DETERMINATION OF RHEOLOGICAL PARAMETERS

4.3.1 *Coefficient of viscosity*

1. *Chu Junda's (1980,1983) model*. Chu starts with the famous Einstein's viscos-

ity equation, which is suitable for the fluid with low concentration of uniform, cohesionless spheres in suspension:

$$\frac{\eta}{\mu_0} = 1 + 2.5\,S_v \tag{4.8}$$

where η and μ_0 are viscosity of sediment-water mixture and that of clear water at the same temperature, S_v is the volumetric concentration.

It is known in physico-chemistry that bound water film is formed around the sediment particles in water. Bound water film firmly adheres to sediment particles and should be regarded as a part of the particles. In other words, the existence of the bound water film increases the effective concentration to a certain degree. Let K stands for the ratio of total volume of particles together with bound water film to the volume of particles. Then the Einstein's viscosity equation may be modified as (Chu, 1980, 1983)

$$\frac{\eta}{\mu_0} = 1 + 2.5\,KS_v \tag{4.9}$$

Let D the diameter of uniform spheres, and δ the thickness of the bound water film. Neglecting the high order of δ/D, one obtains

$$K = [(D + 2\delta)\,/D]^3 = 1 + 6\delta/D \tag{4.10}$$

For non-uniform sediment particles at high concentration, it is assumed that Einstein's equation is still valid for each small concentration increment S_{vi}. Divide all the sediment particles into n groups according to their sizes. Median diameters of these groups are $D_1, D_2, ..., D_i, ..., D_n$, in which $D_{i-1} < D_i$. Each group can be approximately regarded as uniform particles. At first let us put D_1 group of sediment particles into the water, they replace water with the same volume as that of the particles and a suspension at concentration of S_{v1} is formed. The viscosity of the suspension η, can be described by Einstein's equation. Considering the suspension as a new medium, then we put D_2 group into the suspension in the same procedure as before. It is assumed the correlationship between η_1 and η_2 still obeys Einstein's equation. Repeat the same procedure for i times, we got:

$$\frac{\eta_i}{\eta_{i-1}} = 1 + 2.5\,K_i\,S_{vi} \tag{4.11}$$

where η_{i-1} and η_i are viscosities before and after adding D_i group of sediment particles, which replace water with the same volume, respectively. Hence

$$\Delta\eta_i = \eta_i - \eta_{i-1} = 2.5\,\eta_{i-1}K_i\,S_{vi} \tag{4.12}$$

in which K_i is the constant K (Equation 4.10) for ith group of sediment particles.

In fact, not only bound water film adheres to sediment particles but a certain amount of free water is also entrapped in the pores formed by collision particles.

This part of enclosed free water also moves together with solid particles when undergoing shear and increases the effective concentration. Let θ stand for the ratio of the total volume of particles, bound water and enclosed free water to the sum of the first two. The effective concentration of unit volume of muddy water, in which the ith group of particles has been added, can be written as:

$$(\theta K S_v)_i = (\theta K S_v)_{i-1} (1 - K_i S_{vi}) + K_i S_{vi}$$

and

$$\Delta (\theta KS_v)_i = (\theta KS_v)_i - (\theta KS_v)_{i-1} = K_i S_{vi} (1 - (\theta KS_v)_{i-1}) \qquad (4.13)$$

Dividing Equation (4.12) by Equation (4.13) and infinitively dividing the particle groups, we got:

$$\frac{d\eta}{\eta} = \frac{2.5d(\theta KS_v)}{(1 - (\theta KS_v))}$$

Integrating the above equation, one obtained:

$$\frac{\eta}{\mu_0} = (1 - \theta KS_v)^{-2.5} \qquad (4.14)$$

Equation (4.14) shows that $S_v < (\theta k)^{-1} = S_{vm}$; S_{vm} is called as maximum concentration. As the concentration of muddy water reaches the maximum concentration, all the particles contact with each other and form a closely arranged structure. Consequently, η approaches infinity in such case. For closely arranged uniform spheres $\theta = 9/2\pi = 1.4$. For nonuniform particles $\theta = 1.4$ is adopted as first approximation.

Parameter K can be calculated by Equation (4.10) for uniform spheres. For nonuniform particles K can be calculated by the following equation:

$$K = 1 + 6 \int_0^1 \left(\frac{\delta}{D}\right) dp \qquad (4.15)$$

in which dp is the percentage of particles with a diameter D.

Inputting θ and K into Equation (4.14), one got:

$$\eta/\mu_0 = \{1 - 1.4 [1 + 6 \int_0^1 (\delta/D) \, dp] S_v\}^{-2.5} \qquad (4.16)$$

Considering the definition of the maximum concentration S_{vm}, we can rewrite the aforementioned equation as:

$$\eta/\mu_0 = \{1 - S_v/ S_{vm}\}^{-2.5} \qquad (4.17)$$

In practical application Equation (4.14) can be written in finite form as follows:

$$\eta/\mu_0 = \{1 - 1.4 [1 + 6 \sum (\delta/D_i) P_i] S_v\}^{-2.5} \qquad (4.18)$$

in which P_i is the weight percentage of i-group particles.

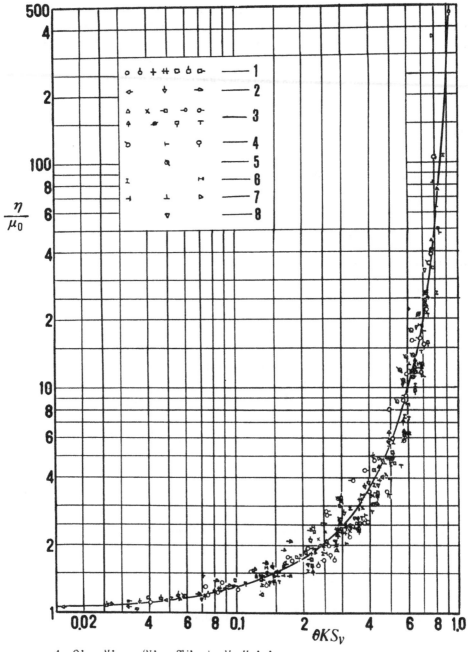

1. Qian Ning (Ning Chien), Ma Huimin
2. Sha Yuqing   3. Ren Zenghai
4. Chi Yaoyu   5. Qian Yiying   6. Chu Junda
7. Williams   8. Vand

Figure 4.3. Comparision between Chu's theoretical curve and experimental data.

The thickness of the bound water film δ can be taken as 1 μm for the sediment after C.N. Woodruff's data, and 0.13 μm for the glass bead after E. Pettijohn's data.

Equation (4.18) has been verified by the author's and others experimental data as Figure 4.3.

It should be pointed out that adoption of $\theta = 1.4$ for non-uniform particles seems unreasonable. For non-uniform particles pores formed by large particles can be filled by small particles and consequently θ value should be smaller.

2. *Fei Xiangjun's method.* Dividing the whole non-uniform sediment particles into several groups with low concentration and nearly uniform composition of particles, adding each group into water-sediment mixture one by one and using Einstein's formula (4.8), Fei developed the expression of relative viscosity for the mixture of water and non-uniform sediment particles as follows (Fei, 1982; Fei & Yang, 1985):

$$\mu_r = \frac{\mu}{\mu_o} = (1 - K'S_v)^{-2.5} \tag{4.19}$$

in which K' is the coefficient of effective concentration.

As mentioned before, in general, the effective concentration of sediment-water mixture consists of three parts: the volume of solid particles V_1, the volume of bound water film around the particles V_2, and the volume of enclosed water V_3, (which varies with sediment size composition and concentration). If D is the diameter of uniform sediment particles, δ is the thickness of bound water film and σ is the porosity of sediment, the coefficient of effective concentration of sediment-water mixture will be given by

$$K' = \frac{V_1 + V_2 + V_3}{V_1} = \left(\frac{D + 2\delta}{D}\right)^3 + \frac{K_1\sigma}{1 - \sigma} \tag{4.20}$$

where K_1 is a coefficient describing the degree of filling of pores by the enclosed water. K_1 is 1 when the pores are fully filled with enclosed water and there is no bound water film, and $K_1 = 0$ when the pores are fully filled with bound water film. As $\delta/D \ll 1$, the terms of higher order can be neglected:

$$K' = 1 + 6\frac{\delta}{D} + \frac{K_1\sigma}{1 - \sigma} \tag{4.21}$$

For non-uniform particle composition with the weight percentage P_i for particle diameter D_i, K' will be:

$$K' = 1 + 6\delta \sum_{i=1}^{n} \frac{P_i}{D_i} + \frac{K_1\sigma}{1 - \sigma} \tag{4.22}$$

It is found that K' is not a constant value, it depends on the volumetric concentration, the grain size composition and the content of fine particles. Fei treated it in two different ways.

1. *Sediment-water mixture without fine particles (D < 0.01 mm)*. In this case, the physico-chemical effect of solid particles with water is very weak, so the contribution of bound water film to the effective concentration of sediment-water mixture can be neglected, and the value of K_1 will approach to unity. The coefficient of effective concentration for uniform sediment becomes:

$$K' = 1 + \frac{\sigma}{1 - \sigma} = \frac{1}{1 - \sigma} \tag{4.23}$$

The porosity of sediment σ depends on the shape and the arrangement of particles.

For uniform spherical particles with prismatical arrangement, the porosity of sediment calculated by geometry is $\sigma = 0.26$. Thus the relative coefficient of viscosity is

$$\mu_r = \left(1 - \frac{S_v}{1 - 0.26}\right)^{-2.5} = (1 - 1.35 \, S_v)^{-2.5} \tag{4.24}$$

It coincides with the expression suggested by Roscoe (1952).

For non-uniform particles with irregular shape, the porosity of sediment may be determined by experiment. The wider the range of sediment size, the smaller the porosity of sediment.

2. *Sediment-water suspension with fine particles*. When there are more fine particles in suspension, for example, the amount of particles with diameter $d < 0.01$ mm is more than 15% by weight, the volume of water film formed around the fine particles due to strong physico-chemical will reach a considerable amount. The coefficient of effective concentration then should be described by Equation (4.21).

As the thickness of water film can not be measured directly, K' can be indirectly determined as follows.

From Equation (4.19), if the volumetric concentration of solids $S_v = 1 / K'$, μ_r will approach infinitive. So S_{vm} may be defined as limiting concentration:

$$S_{vm} = \frac{1}{K'} = \frac{1}{1 + 6\delta \sum\limits_{i=1}^{n} \dfrac{P_i}{D_i}} \tag{4.25}$$

By means of rheological measurements for more than ten samples of sediment-water suspension with different grain size composition at high concentration of solid, an empirical relationship of $S_{vm} \approx 6\sum P_i / D_i$ was established as shown in Figure. 4.4.

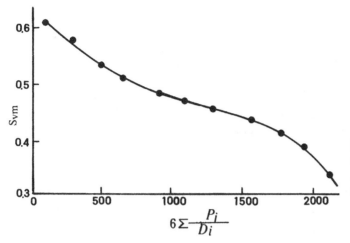

Figure 4.4. The relationship between parameter A and specific surface area (after Fei).

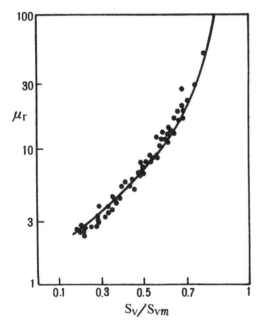

Figure 4.5. $\mu_r - S_v/S_{vm}$.

By virtue of the empirical $S_{vm} \approx 6\sum P_i/D_i$ relationship the limiting concentration S_{vm} of sediment-water suspension with a given size composition can be determined, and then the relative viscosity can be calculated according to Equation (4.19).

As the concentration of suspension S_v is less than 0.35, the following empirical formula is suggested by Fei to take account of the contribution of enclosed water (Fei & Yang, 1985):

$$\mu_r = \left\{ 1 - \left[1 + \frac{3}{2} \left(\frac{S_{vm} - S_v}{S_{vm}} \right)^4 \right] \frac{S_v}{S_{vm}} \right\}^{-2.5} \tag{4.26}$$

The comparison of experimental data with Equation (4.26) is shown in Figure 4.5.

As the whole size composition, instead of a representative diameter, has been taken into consideration, the experimental data for different sediment can thus be unified. Fei did his experiments with samples of loess deposits, so Fei's method is particularly suitable for the Yellow River and its tributaries.

4.3.2 *Bingham yield stress*

1. *The critical concentration for the formation of a Bingham fluid, S_{vo}.* As mentioned above, a slurry becomes a Bingham fluid and Bingham yield stress appears when its concentration reaches a critical value S_{vo}. It can be readily seen from the foregoing discussion that S_{vm} should also be an important parameter in this problem. Actually , the following formula is given by Fei (Fei, 1983):

$$S_{v0} = \beta S_{vm}^{3.2} \tag{4.27}$$

in which $\beta = 1.26$ for sediment slurry; $\beta = 1.87$ for coal slurry.

Considering the practical difficulty in judging the appearance of τ_B, Fei takes $\tau_B = 0.5 \text{N/m}^2$ as the critical condition for distinguishing a Newtonian fluid from a Bingham fluid.

2. *Expressions for Bingham yield stress.* Lots of expressions for Bingham yield stress have been suggested by different authors. Some of them are listed in Table 4.2.

The following concept can be deduced from the table:

1. The yield stress τ_B rapidly increases with concentration increasing.

2. The more the content of fine particles, the larger the yield stress τ_B.

3. The yield stress τ_B of sediment with different mineral components differs greatly. We will discuss it in detail later.

4. Generally speaking, due to the lack of the consideration of the whole size composition and the mineral contents of the sediment, most formulas listed in Table 4.2 can only be applied to the sediment, with which the author did the measurement and developed his formula.

In this respect, Fei's (1983) work got some progress. Using S_{vm} as a parameter reflecting characteristics of the size composition of the sediment, he developed the following empirical formula for yield stress:

$$\tau_B = 9.9 \times 10^2 \exp \left(\xi \frac{S_v - S_{v0}}{S_{vm}} + 1.5 \right) \tag{4.28}$$

where τ_B is in N/m^2, in which $\xi = 8.45$ for sediment-water mixture, and 6.67 for coal-water mixture.

Table 4.2. Expressions for Bingham yield stress.

No	Expression for τ_B (N/m^2)	Sediment	D_{50}(mm)	Authors (year)
1	$\tau_B = 5.88 \times 10^{-3}\, e^{19.1\,S_v}$	Boutou silt	0.0058	Qian (Ning Chien) & Ma (1958)
		Guanting silt	0.0090	
2	$\tau_B = 5.88 \times 10^{-3}\, e^{32.1\,S_v}$	Tanggu silt I	0.0080	
		Tanggu silt	0.0045	
3	$\tau_B = 25.78 \times 10^2\, S_v^{5.4}$	Zhengzhou loess	0.056	Wan et al. (1979)
4	$\tau_B = 98.98 \times 10^2\, S_v^{5.4}$	Zhengzhou loess	0.043	
5	$\tau_B = 282.2 p^{2.23}\, (S \times 10^{-3})^{4.33}$	Silt from Weihe and North Luohe	0.016 ~ 0.07	Zhang et al. (1980)
	$\tau_B = 12.25 p\, (S \times 10^{-3})^{1.72}$			
6	$\tau_B = 255 \times 10^{7.59\,S_v}$	Pinglu loess	0.013	Dai et al. (1980)
7	$\tau_B = 98 \dfrac{S_{vm}}{D} \left\{ 2.5 \left(\dfrac{S_v}{S_{vm}}\right)^{3/2} + 120 \left(\dfrac{S_v}{S_{vm}}\right)^8 \right\}$			Tang (1981)
8	$\tau_B = 1.28 \times 10^3\, S_v$	Kaoline bentonite		Wan (1982)
9	$\tau_B = 4.15 \times 10^5\, S_v$			
10	$\tau_B = 9 \times 10^3 p_1^{1.82}\, S_v$	Huayuankou clay and Qinghuandao sand	0.005	Wang & Qian (Ning Chien) (1984)
11	$\tau_B = 0.225\, e^{18\,S_w}$	Clay		Babbitt & Caldwell (1940)
12	$\tau_B = 1637\, S_v^3$	Kaoline		Thomas (1961)
13	$\tau_B = K_1\, S_v^{(4\sim5)}$	Deposits in bay, reservoir, kaolin, limestone		Migniot (1968)
14	$\tau_B + \varepsilon = K_2 S_v$	Clay		Govier & Aziz (1972)

Remarks: 1. S, concentration in kg/m^3; S_w concentration in weight, that is, the ratio of the weight of the sediment to the weight of the turbid water; S_{vm}, maximum concentration, in volume, of the deposits of suspension. 2. k_1, k_2, constants, ξ, constant. 3. p, the percentage of the particles finer than 0.025 mm. p_1, the percentage of the particles finer than 0.01 mm.

The comparison of the experimental data and Equation (4.28) is shown in Figure 4.6. The agreement between them is quite satisfied.

3. *Bingham yield stress of hyperconcentrated flow of coarse particles.* Does Bingham yield stress exist in the suspension purely consisting of particles coarser than 0.01 mm?

It is a common understanding that suspension purely consisting of coarse

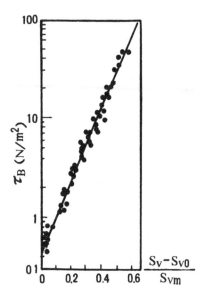

Figure 4.6. $\tau_B - \dfrac{S_v - S_{v0}}{S_{vm}}$.

particles at low concentration behaves as a Newtonian fluid. According to the experiments with Qinghuangdao sand carried out by Wang, the suspension at concentration $S_v < 0.27$ still behaves as a Newtonian fluid. It can be approximatively described by a pseudo-plastic model if the concentration is higher (Wang, Z., 1984). Based on Wang Xinsheng's (1981) experimental results, Bingham yield stress appears as the concentration of a suspension purely consisting of coarse particles (> 0.01 mm) surpasses 0.35. In this case the yield stress is a reflection of the frictional resistance due to mutual collision between particles, and its absolute value is much smaller than that in case of fine particles. For smooth spherical plastic beads, the friction is small and S_v may reach 60%.

4. *Bingham yield stress of suspension consisiting of both fine and coarse particles.* Shen & Xie (1983) suggested a model to treat the Bingham yield stress of suspension consisting of both fine and coarse particles. It is considered that fine particles and water form a homogeneous fluid and coarse particles disperse in the homogeneous fluid. Shear stress consists of two parts, viscous force and dispersive force. Viscous force can be written as:

$$\tau' = \tau_B + \eta \frac{du}{dy}$$

Considering the fact that the space occupied by coarse particles can not deform during the motion and consequently the shear rate in the rest part of the space occupied by fluid increases to a certain degree, the authors introduced a factor $(1 + (0.25\,\pi\lambda^3)/(1 + \lambda)^2)$ to consider this effect.

$$\tau' = \tau_B + \left[\frac{\pi}{4}\frac{\lambda^3}{(1 + \lambda)^2}\right]\eta\frac{du}{dy} \tag{4.29}$$

Dispersive force τ'' can be described by Bagnold's formula in principle, but the authors did some revision as follows:

$$\tau'' = K\lambda^n\eta\frac{du}{dy} \tag{4.30}$$

in which λ is the linear concentration defined as $\lambda = D/s$, s the interval between two neighbouring particles. Based on experimental data, they got the coefficients $n = 2$ and $K = 0.09$.

They wrote down the whole shear stress as:

$$\tau = \left\{\left[1 + \frac{\pi}{4}\frac{\lambda^3}{(1 + \lambda)^2}\right] + K\lambda^n\right\}\eta\frac{du}{dy} \tag{4.31}$$

It should be pointed out that:

1. The authors omitted the yield stress τ_B in the deduction of Equation (4.31). It seems unreasonable.

2. When the authors considered the increases of shear rate due to the existence of coarse particles, they treated the three dimensional problem as a two dimensional one. So further study is needed.

As a first approximation in practise, the following treatment can be recom-

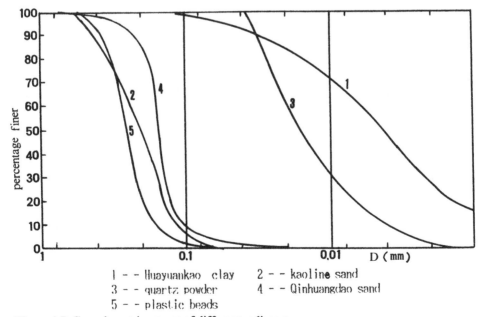

1 – – Huayuankao clay   2 – – kaoline sand
3 – – quartz powder   4 – – Qinhuangdao sand
5 – – plastic beads

Figure 4.7. Granulometric curves of different sediment.

manded. Considering the fact that coarse particles have little effect on yield stress, we can exclude coarse particles (for instance, particles coarser than 0.10 mm or 0.15 mm) from the mixture and measure the yield stress of the residual part, and take the measured yield stress as the yield stress of the whole mixture. Practise proves that such treatment is applicable.

5. *The effect of mineral contents on Bingham yield stress.* As mentioned above, the magnitude of Bingham yield stress of a slurry depends not only on concentration S_v and the size composition of the sediment, but also on its mineral content. As listed in Table 4.2, at same concentration S_v, the yield stress of a bentonite slurry is about 300 times larger than that of a kaoline slurry (Wan, 1982). Z. Wang also compared rheological properties of slurry containing different mineral contents. The granulometric curves of different sediment are shown in Figure 4.7. The corresponding rheograms of some sediment are shown in Figure 4.8.

Although quartz powder is much finer than kaoline, but slurry of quartz

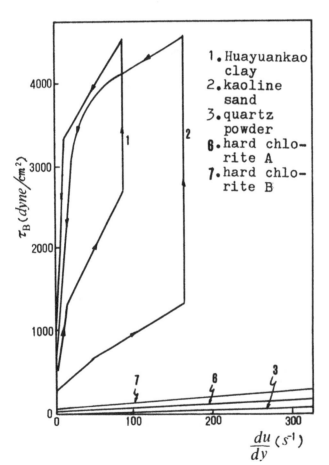

1. Huayuankao clay
2. kaoline sand
3. quartz powder
6. hard chlorite A
7. hard chlorite B

Figure 4.8. Rheograms of suspensions consisting of particles with different mineral composition.

powder-water mixture at $S_v = 0.33$ still behaves as a typical Newtonian fluid, and its viscosity is not very high. In the meantime, the slurry of kaoline containing no particles finer than 0.01 mm at $S_v = 0.32$ behaves as non-Newtonian fluid with rather high Bingham yield stress and rigidity.

4.4 THE INFLUENCE OF TURBULENCE ON BINGHAM YIELD STRESS

In hyperconcentrated fluid flocculation structure is formed due to the surface physico-chemical action of fine particles. And the flocculation structure results in Bingham yield stress. As mentioned above, owing to the existence of Bingham yield stress, particles with a certain diameter D can be supported in still fluid and do not settle. The diameter D is proportional to the Bingham yield stress.

Experiments carried out by Song et al. (1986) clearly shows the effect of turbulence on flocculation structure and consequently the yield stress (Song et al., 1986). Experiments were carried out with sand-clay-water mixture. At first let it run with high velocity and all the sand particles were fully suspended. If the valve was suddenly closed and the flow stopped, no coarse particle deposited on the bottom of pipe even after more than 20 hours. Then if we opened the valve and increased discharge gradually, deposition appeared on the bottom. The thickness of the deposition increased with velocity increasing in the first stage, see Figure 4.9. After the thickness of deposition reached a maximum value at point A, the further increase of velocity made the deposition thickness decrease gradually and disappear at last.

The explanation of above mentioned phenomena is as follows. In a flowing suspension turbulence destroys part of the flocculation structure of the Bingham fluid. It results in the reduction of Bingham yield stress. Consequently, particles, which can be supported by Bingham yield stress and do not settle in stationary

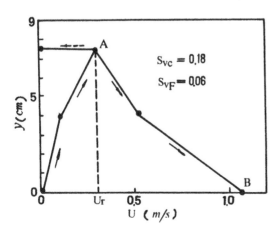

Figure 4.9. The variation of deposition thickness with velocity. S_{vc} and S_{vF} are concentration of coarse particles and that of clay particles, respectively.

Bingham fluid, may deposit in a flowing Bingham fluid. In the first stage ($U < U_r$ in Figure 4.9) the flocculation structure is destroyed more and more as the velocity increases. Correspondingly, the deposition layer becomes thicker and thicker. As the velocity is higher than U_r, although the flocculation structure is further destroyed with velocity increasing, turbulence already plays a leading role. More particles can be entrained and suspended as velocity increases. As a result, the deposition layer becomes thinner and thinner.

Yang & Chien tried to quantitatively describe this phenomenon and got the similar conclusion (Yang & Chien, 1986).

They adopted the simplified model suggested by F. Moore to describe the formation and development of the flocculation structure. It is imagined that the flocculation structure is formed through numerous 'links'. Links which have not connected are called as free links. Those which have connected are called as structural links. The formation of flocculation structure is just the course of the connection of free links and their transforming into structural links. Hence, the ratio of the amount of structural links to the total amount of all the links, which is denoted by λ can be used to reflect the degree of flocculation. $\lambda = 0$ means no flocculation structure at all. $\lambda = 1$ means that all the links have transformed into structural links.

The flocculation structure is in a dynamic equilibrium. Some free links get connection with others and transform into structural ones when they touch each others. Links are fragile and easy to be broken. Some structural links may be detached and transform into free links. Under certain condition the connection of free links and the detachment of structural links reach equilibrium and keep a certain value. The λ value of a slurry at a certain concentration S_v under stationary condition is denoted by λ_c. The corresponding λ value under turbulent condition is λ_T. Then the ratio $Y = \lambda_T / \lambda_c$ is called as the coefficient of turbulence effect. It can be used to reflect the effect of turbulence on flocculation structure.

Yang did a series of experiments in a cylindrical container, in which a uniform

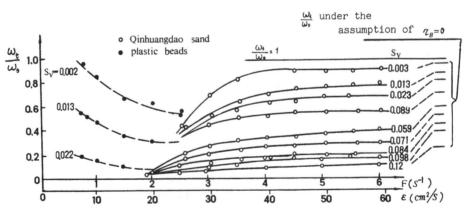

Figure 4.10. The variation of ω_t / ω_0 with turbulence intensity.

turbulence field with different intensities is generated by a set of oscillating grids. At first, the concentration profiles of coarse particles in clear water under different turbulence intensities were measured. Based on these data the correlationship between the momentum transfer coefficient ε, which is used to characterize the intensity of turbulence, and the frequency F of oscillating grids was estabtished. Then the concentration profiles of coarse particles in slurry with different concentration were measured. As the concentrations of slurry were not too high, it is reasonable to assume that the ε in clear water equals to the corresponding one in slurry. So the fall velocity ω_t of coarse particles in slurry under different ε could be deduced from the concentration profiles. The variation of ω_t/ω_0 with ε increasing is shown in Figure 4.10. Here ω_0 is the fall velocity of coarse particles in clear water. The explanation of such tendency is as follows. Under the condition of low concentration and small ε, turbulence increases the collision of particles and promotes the formation of flocculation structure. As a result, the fall velocity of coarse particles decreases. In the region of intensive turbulence the effect of turbulence on breaking structural links plays a leading role. And the flocculation structure deteriorates with increasing ε. Consequently, the fall velocity of coarse particles ω_t increases.

Yang et al. also got the variation of the coefficient of turbulence effect Y, η_t/η, τ_{BT}/τ_B with ε as shown in Figure 4.11. In which τ_B and η are the Bingham yield stress and the rigidity of the slurry under the condition of flow without turbulence, correspondingly. And τ_{BT} and η_t are the Bingham yield stress and the rigidity of the slurry under the turbulence condition. Details of the deduction can be found in the reference.

Jiang also studied the effect of turbulence on flocculation structure and Bingham yield stress of a slurry (Jiang, 1987). Experiments were also carried out in a rectangular container, in which a uniform turbulence field was generated by a set of horizontally oscillating grids. The container was filled with water at first, and then slurry with different concentrations. Coarse particles treated by radioisotope were put into the water or slurry and their fall velocities were directly measured by sensors. Based on the measured fall velocities at different turbulence intensities, the variation of the flocculation structure and the corresponding variation of Bingham yield stress could be deduced. Main conclusions of this study are as follows:

1. Particles, which do not settle or extremely slowly settle in stationary slurry, settle with a certain velocity when grids oscillate and turbulence is generated. The higher the slurry concentration, the more obvious the difference.

2. The smaller the Reynolds number $\omega_0 D/\nu$, the larger the ω_t/ω, see Figure 4.12. In which D and ω_0 are the diameter of coarse particles and their fall velocity in clear water, respectively, ω_t and ω are the fall velocity of coarse particles in slurry with turbulence and that in slurry without turbulence, respectively.

3. The increase of the fall velocity is caused by the reduction of yield stress τ_{BT} in turbulent slurry. Turbulence destroys the flocculent structure and makes yield

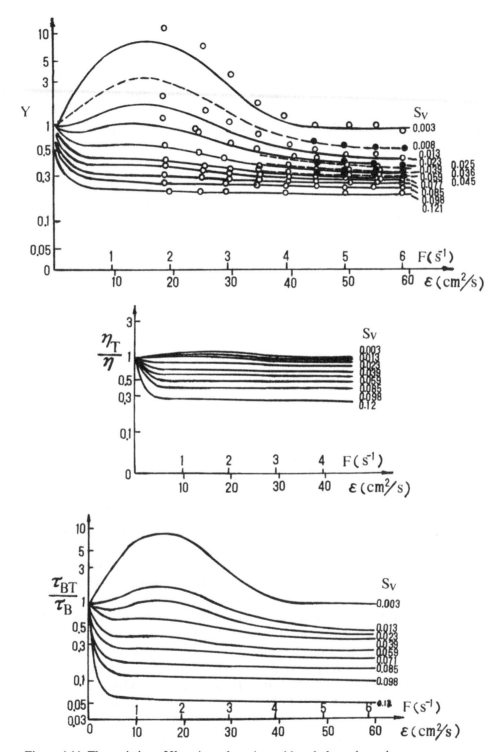

Figure 4.11. The variation of Y, η_T/η and τ_{BT}/τ_B with turbulence intensity.

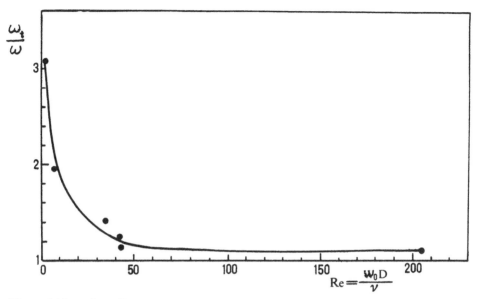

Figure 4.12. $\omega_t / \omega \sim R_{\acute{e}}$.

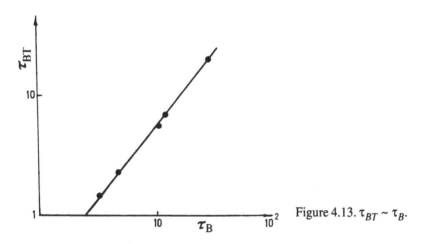

Figure 4.13. $\tau_{BT} \sim \tau_B$.

stress reduce correspondingly. In a fully developed turbulent slurry the yield stress τ_{BT} can be described by the following equation, see Figure 4.13:

$$\tau_{BT} = 0.372 \, \tau_B^{1.167} \tag{4.32}$$

in which τ_B is the yield stress obtained by rheological measurement. In the mean time, in a fully developed turbulent slurry rigidity η reduces only a little.

In Wan's (1982) rheological measurement with rotating viscometer bentonite suspension showed thixotropy characteristics, see Figure 4.14. In Wang's (1982)

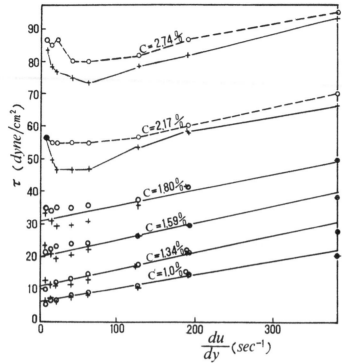

Figure 4.14. Unusual rheological phenomena.

+ under the condition of increasing
  shear rate
₀ under the condition of decreasing
  shear rate

rheological measurement also with rotating viscometer some unusual phenomena have been observed, see Figure 4.8. For Huayuankao clay slurry and kaoline slurry at hyperconcentration shear stress rapidly increases under the condition of keeping shear rate constant. They also behaved as thixotropic fluid.

There are many other examples showing the complex properties of slurry at hyperconcentration.

In general cases one may expect the shear stress within a fluid to be related not only to the shear rate but to all factors which determine the concentration and resistance of particles or particle agglomerates (Govier & Aziz, 1972). Such factors include the formation or destruction of agglomerates, the alignment or disalignment of particles or agglomerates in the direction of the shear stress. Viscoelastic effects and time-dependent properties may be important in sudden changes in flow rate (start-up and stop), in rapid oscillatory flows, and in flows where changes of cross section are encountered. On the other hand, in most normal steady-state flow, and in unsteady-state flow involving long process times, the formation or destruction of agglomerates, the alignment or disalignment of materials behave essentially as purely viscous material and can be described by simpler models.

REFERENCES

Babbitt, D.E. & B.J. Glidden 1940. Turbulent flow of sludges in pipes. *Bull., University Illinois*, Vol. 38(13): pp. 44.

Bird, R.B., G.C. Dai & B.J. Yarusso 1982. The rheology and flow of viscoplastic materials. *Reviews in Chemical Engineering*, Vol. 1(1): 1-70

Chu, J. 1980. The viscosity of sediment-water mixture (in Chinese). *Proc. of The International Symposium on River Sedimentation*, Vol. 1: 205-212.

Chu, J. 1983. Basic characteristics of sediment-water mixture with hyperconcentration (in Chinese). *Proc. of the Second International Symposium on River Sedimentation*, pp. 256-273.

Dai, J., Z. Wan, W. Wang, W. Chen & X. Li 1980. An experimental study of slurry transport in pipes (in Chinese). *Proc. of the International Symposium on River Sedimentation*, Vol. 1: 195-204.

Fei, X. 1982. Viscosity of the fluid with hyperconcentration-coefficient of rigidity (in Chinese). *Journal of Hydraulic Engineering*, No.3: 57-63.

Fei, X. 1983. Grain composition and flow properties of heavily concentrated suspension (in Chinese). *Proc. of the Second International Symposium on River Sedimentation*, pp. 307-308.

Fei, X. & M. Yang 1985. The physical properties of flow with hyperconcentration of sediment. *Proc. of International Workshop on Flow at Hyperconcentrations of Sediment*, pp. 2.

Govier, G.W. & K. Aziz 1972. The flow of complex mixtures in pipes. *Van Nostrand Reinhold*, pp. 24-26, 49.

Jiang, N. 1987. Experiment on fall velocity of spheres in turbulent slurry (in Chinese). *Report of Institute of Hydraulic Research, Yellow River Conservancy Commission*, pp. 29.

Migniot, C. 1968. A study of the physical properties of various forms of very fine sediment and their behaviour under hydrodynamic action. *La Houille Blanch*, No.7.

Pazwash, H. & J.M. Robertson. Fluid-dynamic consideration of bottom materials. *Proc. ASCE*, Vol. 97(HY9): 1317-1329.

Qian, N. (Ning Chien) & H. Ma 1958. Viscosity and rheogram of muddy water (in Chinese). *Sediment Research*, Vol. 3(3): 52-77.

Qian, N. (Ning Chien) & Z. Wan 1986. A critical review of the research on the the hyperconcentrated flow in China. *Series of Publication of International Research and Training Centre on Erosin and Sedimentation*, pp. 42.

Roscoe, R. 1952. The viscosity of suspensions of rigid spheres. *Br. J. of Applied Physics*, Vol. 3: 267-269.

Sha, Y. 1947. Discussion on sediment category and its nomenclature (in Chinese). *Water Conservancy*, Vol. 15(1): 152-178.

Sheng, S. & S. Xie 1983. Mode of structure of debris flow and the effect of coarse grains on the rheological characteristics of slurry (in Chinese). *Journal of Sediment Research*, No.3: 12-19.

Song, T., Z. Wan & N. Qian 1986. The effect of fine particles on the two-phase flow with hyperconcentration of coarse grains (in Chinese). *Journal of Hydraulic Engineering*, No.4: 1-10.

Tang, C. 1981. Formula for the Bingham limit shear stress in sediment-water mixture (in Chinese). *Journal of Sediment Research*, No.2: 60-65.

Thomas, D.G. 1961. Transport characteristics of suspensions, laminar flow properties of flocculated suspensions. *J. Amer. Inst. Chem. Engrs*, Vol. 7(3): 431-437.

Wan, Z. 1982. Bed material movement in hyperconcentrated flow. *Ser. Paper 31, Inst. of Hydrodynamics and Hydraulic Engineering, Tech. Univ. of Denmark*, 79 pp.

Wan, Z., Y. Qian, W. Yang & W. Zhao 1979. An experimental study on hyperconcentrated flow (in Chinese). *People's Yellow River*, No.1: 53-65.

Wang, M., Y. Zhan, J. Liu, W. Duan & W. Wu 1983. An experimental study on turbulence characteristics of flow with hyperconcentration of sediment. *Proc. of the Second International Symposium on River Sedimentation*, pp. 36-46.

Wang, X. 1981. Rheological properties of slurry without clay particles. *Thesis of Tsinghua University.*

Wang, Y. 1982. The rheological properties of debris flow (in Chinese). *Journal of Sediment Research*, No.2: 74.78.

Wang, Z. 1984. Experimental study on the mechanism of hyperconcentrated flow. Ph.D Thesis, Institute of Water Conservancy and Hydroelectric Power Research.

Wasp, E.J. 1977. Solid-liquid flow slurry pipeline transportation. *Trans. Tech. Publications, Germany*.

Yang, M. & N. Qian (Ning Chien) 1986. The effect of turbulence on the flocculent structure of the slurry of fine grains (in Chinese). *Journal of Hydraulic Engineering*, No.8: 30-35.

Zhang, H., Z. Ren, S. Jiang, D. Sun & N. Lu 1980. Settling of sediment and the resistance to flow at hyperconcentrations (in Chinese). *Proc. of the International Symposium on River Sedimentation*, pp. 185-192.

CHAPTER 5

Fall velocity of sediment particles in hyperconcentrated flow

Fall velocity of sediment particles is an important parameter in sediment transport. The fall velocity of sediment particles in hyperconcentrated flow may be several orders of differences from the fall velocity of a single particle in water at rest at infinity.

Patterns of particle settling in hyperconcentrated flow are quite different in different cases. For cohesionless (coarse) particles they settle separately and the mutual influence between them is purely hydrodynamical. This problem will be discussed in Sections 6.1-6.4. For cohesive clay particles at low concentration flocs are formed due to flocculation. In such case, instead of the fall of individual fine particles, discrete flocs settle together with discrete coarse particles. As the concentration of clay particles reaches a critical value, a flocculent network is formed and discrete coarse particles settle in it. With further increase of the concentration of clay particles, the flocculent network makes all the particles settle together at a uniform but very low velocity. These problems as mentioned above are the topics of Sections 5.5 to 5.8. At last, some unusual phenomena associated with extremely high concentration of clay particles will be discussed in Section 5.9.

5.1 FACTORS CAUSING REDUCTION OF FALL VELOCITY

Fall velocity may reduce because of several influences. Wan & Sheng (1978) illustrated it through the following deduction:

Let us consider a single particle settling in water at rest at infinity. The particle reaches its ultimate constant fall velocity as soon as its submerged weight is balanced by a drag force, which can be described by Stokes' law in laminar flow region, that is:

$$(\gamma_s - \gamma_0)\frac{\pi}{6}D^3 = 3\pi\mu_0 D\omega_0 \tag{5.1}$$

in which ω_0 is the fall velocity of a single particle with diameter D, μ_0 the viscosity of clear water, γ_s and γ_0 the specific weight of the sediment particle and that of the water, respectively.

For group fall velocity of uniform sediment particles in a suspension with concentration S, some changes have to be taken into consideration.

First, based on the consideration of continuity, the settling of particles is bound to induce an upwater flow with a velocity u.

$$u = \frac{S_v}{1 - S_v} \cdot \omega$$

where ω is the group fall velocity and is smaller than ω_0 in Equation (5.1).

The relative velocity between settling particles and upward flow is the sum of the fall velocity ω and the upward velocity of the surrounding fluid u.

$$u + \omega = \frac{\omega}{1 - S_v} \tag{5.2}$$

It is the relative velocity which causes the drag force.

Secondly, the specific weight γ_m of the suspension with concentration S_v is different from that of the clear water γ_0.

$$\gamma_m = \gamma_0 + (\gamma_s - \gamma_0)\, S_v$$

and

$$\gamma_s - \gamma_m = (\gamma_s - \gamma_0)\,(1 - S_v) \tag{5.3}$$

The submerged weight of the considered particle is reduced due to the increase of the specific weight of ambient fluid and the consequent increase of the buoyance force, $\gamma_m\, \pi/6\, D^3$.

Finally, the dynamic viscosity increases due to the existence of solid particles. It can be considered by introducing Moliboxino's formula (1956).

$$\frac{\mu}{\mu_0} = 1 + \frac{3}{\dfrac{1}{S_v} - \dfrac{1}{0.52}} \tag{5.4}$$

in which μ is the viscosity of suspension with concentration S_v.

Substituting all these above mentioned modifications into Equation (5.1), one obtains:

$$(\gamma_s - \gamma_m)\frac{\pi}{6}D^3 = 3\pi\mu D\, \frac{\omega}{1 - S_v}$$

or

$$(\gamma_s - \gamma_0)\frac{\pi}{6}D^3 = 3\pi\mu D\, \frac{\omega}{(1 - S_v)^2} \tag{5.5}$$

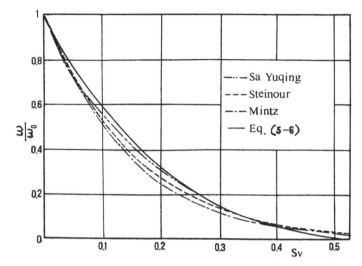

Figure 5.1. $\omega/\omega_0 \sim S_v$.

Dividing the foregoing formula by Equation (5.1) and substituting Equation (5.4), one obtains:

$$\frac{\omega}{\omega_0} = \frac{(1 - S_v)^2}{1 + \dfrac{3}{\dfrac{1}{S_v} - \dfrac{1}{0.52}}} \tag{5.6}$$

Among the three effects mentioned above, the increase of viscosity plays the most important role in determining the particle fall velocity.

The comparision of Equation (5.6) with existing formulas is shown in Figure 5.1. The agreement is quite satisfied.

Based on Stokes' law, Hawksley (1951) also considered the increase of the viscosity and specific weight and the backflow caused by the settling of particles. He deduced the following formula:

$$\frac{\omega}{\omega_0} = \xi(1 - S_v)^2 \exp\left(\frac{-k_1 S_v}{1 - k_2 S_v}\right) \tag{5.7}$$

in which $\xi = 1$ for sediment particles without flocculation and $\frac{2}{3}$ for sediment particles of sphere form and with flocculation, K_1 is a coefficient pertaining to the influence of particle shape. $K_1 = \frac{5}{2}$ for spheres, and $\frac{5}{2} \wedge$ for other particles. $\wedge$ is the spherity, K_2 is a coefficient considering the interaction between particles and is equal to $\frac{39}{64}$ for spheres.

Steinour (1944) also considered the increase of viscosity and the specific weight and the backflow and obtained the following formula:

$$\frac{\omega}{\omega_0} = [1 - (1 + \varepsilon) S_v]^2 10^{-1.82(1 + \varepsilon)S_v} \tag{5.8}$$

Table 5.1. Values of ε.

Sediment	D (mm)	Flocculation	ε
Glass spheres	0.0135	Without flocculation	0
Carbonrundum emery	0.0096-0.0122	Without flocculation	0.200
Emery	0.0122	With flocculation	0.366
	0.0096		0.404
	0.0046		0.538

He considered that because of corners and uneven surface of particles water molecules attach on particles and move together with settling sediment particles. Attached water molecules actually increases the effective sediment concentration. The volume of the attached water is assumed to be proportional to the volume of sediment particles and the ratio is ε. The values of ε are large for flocculated fine particles. The ε values for different sediment particles are listed in Table 5.1.

5.2 BUOYANCY FORCE OF A BODY IN SOLID-LIQUID MIXTURE

The submerged weight of a particle is the difference of its weight $\gamma_s \, \pi/6 \, D^3$ and the buoyance force acting on it F. In the foregoing section the buoyancy force F is calculated by the following equation.

$$F = \gamma_m \frac{\pi}{6} D^3 \tag{5.9}$$

To test Equation (5.7), Wang (1987) carried out a series of tests with an experimental set-up shown in Figure 5.2. The vessel was filled with mixture consisting of homogeneous quartz sand and water or homogeneous sand and glycerine. Glycerine was of very high viscosity so that it could prevent suspended quartz sand from falling quickly. Quartz sand of different sizes was used in his experiments. A copper ball with a weight of 1.52 grams was put into mixture consisting of glycerine/water-quartz sand and was hanged from an electronic scale with a length of thin thread. The buoyancy force acting on the copper ball can be given by the weight of the ball minus the reading shown on the electronic scale. The reading was taken when the quartz particles suspended all around the copper ball and the volumetric concentration of suspended sand kept constant. The results of the tests are listed in Table 5.2, in which D_m is the diameter of the copper ball (6.9 mm) and D is that of quartz sand. And ΔF is defined as:

$$\Delta F = F - \gamma_0 \frac{\pi}{6} D_m^3 \tag{5.9}$$

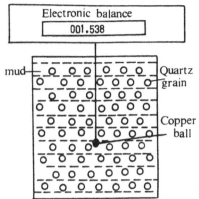

Figure 5.2. Experimental set-up.

Table 5.2. The variation of ΔF and $f(D_m/D)$.

Liquid phase	γ_0 (g/cm^3)	γ_s (g/cm^3)	D (mm)	S_v	F (g)	ΔF (g)	D_m/D	$f(D_m/D)$
Glycerine	1.22	2.65	0.90	0.36	0.265	0.058	7.67	0.71
Glycerine	1.22	2.65	0.90	0.33	0.260	0.053	7.67	0.66
Glycerine	1.22	2.65	0.47	0.29	0.260	0.053	14.7	0.76
Glycerine	1.22	2.65	0.11	0.294	0.275	0.068	62.7	0.95
Water	1.00	2.65	0.015	0.382	0.280	0.110	460.0	1.01

where ΔF represents the increment of the buoyancy due to the existence of the suspended sand.

As indicated by Table 5.2, ΔF increases with increasing D_m/D.

In case the ball has the same size as that of the suspended sand, that is, $D_m/D = 1$, by virtue of Stokes equation and the distribution formula of fluid pressure around a spherical particle moving relative to ambient liquid, Wang (1987) derived buoyancy as follows:

$$F = \gamma_0 \frac{\pi}{6} D_m^3 + 0.134 \, (\gamma_s - \gamma_0) \, S_v \frac{\pi}{6} D_m^3 \tag{5.10}$$

or

$$\Delta F = 0.134 \, (\gamma_s - \gamma_0) \, S_v \frac{\pi}{6} D_m^3 \tag{5.11}$$

Generally speaking, ΔF can be rewritten as:

$$\Delta F = f\left(\frac{D_m}{D}\right) \cdot (\gamma_s - \gamma_0) \, S_v \frac{\pi}{6} D_m^3 \tag{5.12}$$

Using data in Table 5.2 together with Equation (5.11). Wang got the function $f(D_m/D)$ as shown in Figure 5.3.

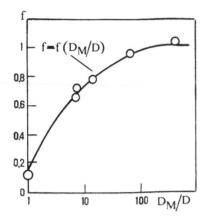

Figure 5.3. Coefficient f versus relative size D_M/D.

The buoyancy increases with increasing D_m/D. $f(D_m/D)$ approaches 1 when D_m/D is greater than about 50.

5.3 GROUP FALL VELOCITY OF UNIFORM COHESIONLESS SEDIMENT PARTICLES

Due to the interaction between particles, which will be discussed in detail later, the fall velocity of cohesionless coarse particles is smaller than that of a single particle. The collective fall velocity of coarse particles is named as group fall velocity.

The most widely adopted formula of the gross fall velocity of uniform discrete sediment particles is suggested by Richardson & Zaki (1954).

$$\frac{\omega}{\omega_0} = (1 - S_v)^m \tag{5.13}$$

in which ω is the gross fall velocity of uniform discrete sediment particles in solid-liquid mixture with concentration S_v, ω_0 the fall velocity of a single particle in pure liquid. This formula is simple in form and convenient to be used.

Qian (1980) suggested that the exponent m is a function of grain Reynolds number.

$$m = f\left(\frac{\omega_0 D}{\nu}\right) \tag{5.14}$$

in which ν is the kinematic viscosity of clear water. Based on experimental data, this function of grain Reynolds number is given as Figure 5.4. m approaches a maximum constant value 4.65 as the Reynolds number is smaller than 0.4, and approaches a minimum constant value 2.5 when the Reynolds number surpasses $10^3 - 10^4$, as shown in Figure 5.4.

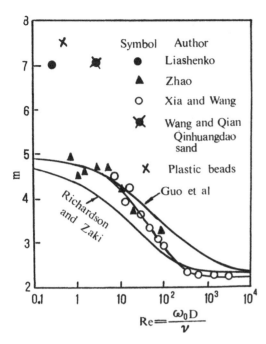

Figure 5.4. $m \sim \omega_0 D / \nu$.

Later Xia and Wang conducted experiments with uniform noncohesive sand ($D_{50} = 0.067$ mm) and obtained $m = 7$, which is larger than the maximum value 4.65 given by Qian. Wang and Qian did settling experiments with uniform sand ($D_{50} = 0.15$ mm) and obtained $m = 7$.

5.4 GROSS FALL VELOCITY OF NON-UNIFORM COHESIONLESS PARTICLES

Qian et al. (1980) conducted experiments with non-uniform cohesionless particles of D_{50} 0.061 mm and 0.087 mm. The experiment results are shown in Figure 5.5. Fall velocities of non-uniform cohesionless particles at different concentrations are plotted by dots and circles. They are distributed just along the curve for uniform discrete particles. It means that non-uniform discrete particles follow the same law as that for uniform dicrete particles.

5.5 SETTLING OF SEDIMENT CONTAINING CLAY PARTICLES

Settling phenomena of sediment containing clay particles is much more complicated. Figure 5.6 shows the average fall velocity of two kinds of sediment consisting of clay and silt.

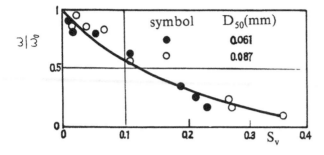

Figure 5.5. The fall velocity of non-uniform discrete particles.

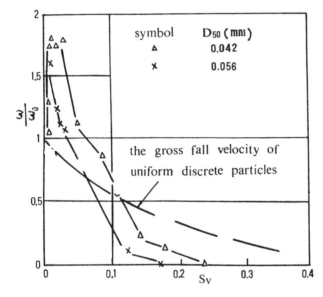

Figure 5.6. The fall velocity of sediment containing clay particles under different concentration.

The sediment with D_{50} = 0.042 mm had clay (< 0.005 mm) content of 14%. At low concentration, for instance, S_v < 0.026, the fall velocity increases with concentration increasing, and the fall velocity of non-uniform sediment is larger than the fall velocity of a single particle with a diameter of 0.042 mm settling in water at rest at infinity.

The explanation is as follows. Due to the action of flocculation, fine clay particles connect to each other during random collision and form flocs, which have larger diameter than individual clay particles. Consequently, flocs fall faster than individual particles. As concentration increases, the opportunity of random collision increases and more flocs are formed. As a result, the fall velocity increases.

With concentration further increasing, flocs connect to each other and a continuous flocculent network is formed in the whole vessel. Such flocculent network has a certain structural strength and hence the fall velocity rapidly drops down. At the beginning only clay particles form the flocculent network. Actually, these clay particles together with water form homogeneous slurry, which settles extremely slowly in the form of a descending interface between clear water and turbid water. Maintaining discrete group, coarse particles settle through the flocculent structure and hence their fall velocity is greatly deduced. Sorting phenomenon still exists during such settling.

As concentration further increases, more and more coarse particles are entrapped into the flocculent structure, and the sorting phenomenon becomes less and less obvious. Finally, all the particles are entrapped into the flocculent structure and the mixture becomes a homogeneous fluid.

The foregoing development is clearly revealed in an experiment carried out at the Shaanxi Provincial Institute of Hydraulic Research (1977). In the experiment non-uniform sediment with D_{50} = 0.035 mm was used and the original concentration, that is, the sediment concentration at the beginning of the experiments, varied from 0.075 to 0.302.

At low original concentration, for instance, S_{v0} = 0.0075, coarse particles settled quickly and sorting phenomenon was obvious. Only a few fine particles together with water formed homogeneous fluid with flocculent structure and settled with a fall velocity of 0.00345 cm/s. The concentration of homogeneous fluid was 0.027 and the median diameter of fine particles forming homogeneous fluid was 0.0089 mm. At higher original concentration, more fine particles formed homogeneous fluid and the interface between clear water and turbid water, i.e. the homogeneous fluid, settled more slowly. As the original concentration S_{v0} reached 0.302, all the sediment particles, including the coarsest ones, participated in forming homogeneous fluid. Therefore the homogeneous fluid had the same concentration and the same median diameter as the original mixture did. And the interface between clear water and homogeneous fluid descends with an extremely low velocity 0.00067 cm/s. Results of the experiment are shown in Table 5.3.

Based on the mechanism mentioned above, the settling sediment containing clay particles can be classified into three categories according to the concentration as well as the sediment constituent, which is represented by the concentration for maximum viscosity S_{vm}, as shown in Figure 5.7 (Qian & Wan, 1986). The definition of S_{vm} is given in Chapter 4.

Category I: Settling of discrete particles and discrete flocs. Fine particles form

Table 5.3. Settling processes of non-uniform sediment at different original concentration.

Original concentration S_{vo}		0.075	0.170	0.226	0.302
Properties of fine sediment particles, forming homogeneous fluid	Median diameter D_{50} (mm)	0.0089	0.0122	0.0133	0.035
	Concentration	0.027	0.080	0.112	0.035
Fall velocity of the interface (cm/s)		0.00345	0.00085	0.00083	0.00067

flocs but the flocculent net structure has not fully developed yet. Discrete flocs and discrete coarse particles settle individually, although there is some interference between them.

Category II: Settling of discrete coarse particles in suspension with flocculent structure. Fine particles form flocculent structure and therefore the suspension exhibits Bingham yield stress. While settling in the flocculent network, discrete particles are subjected to a large drag by the latter.

Category III: Slow settling of flocculent network as a whole. All the particles, including the coarsest ones, participate in forming homogeneous fluid and transform into neutral buoyant load. All the particles slowly settle in an entirety. Actually, it is a consolidation process of the deposit at the bottom of the container.

The transition curve A in Figure 5.7 from Category I to Category II is nothing else but the critical condition for forming a Bingham fluid. The division line B in Figure 5.7 between Category II and III corresponds to the transition from a two-phase flow to a one-phase (or homogeneous) flow. Such transition happens at a certain concentration for a certain constituent of sediment. In Category III all the particles transform into neutral buoyant load. The flocculent structure prevents the coarse particles, including the coarsest particles represented by D_{95}, from settling freely and all the particles settle slowly as an entirety. The larger the D_{95}, the higher the concentration at which the transition happens. As an example, transition curves B for $D_{95} = 0.10$ mm and 0.50 mm are plotted respectively in Figure 5.7.

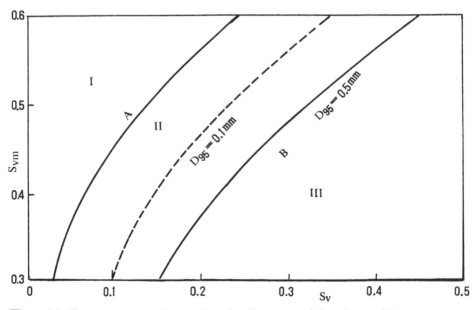

Figure 5.7. Three categories of the settling of sediment containing clay particles.

5.6 SETTLING OF DISCRETE FLOCS AND DISCRETE PARTICLES

As the concentration is lower than the critical concentration defined by line A in Figure 5.7, discrete flocs and discrete particles settle individually.

A. *Fall velocity of discrete coarse particles.* The settling of coarse particle is damped by two factors, namely, the mutual interference between discrete particles and the damping effect caused by flocs on discrete settling particles. Xia & Wang (1982) considered and combined these two factors. A series of experiments, in which uniform sand of 0.067 mm was used as discrete coarse particles and Huayuankou clay was used as fine particles, were carried out by Xia and Wang. Properties of the Huayuankou clay were as follows: $D_{50} = 0.0054$ mm, $D_{95} = 0.0295$ mm, $D_5 = 0.00037$ mm and 71.5% of the clay particles were finer than 0.01 mm.

Let us consider the mutual interference between discrete particles at first:

$$\omega' = \omega_0 (1 - S_{vc})^m \tag{5.15}$$

in which ω_0 = fall velocity of a single particle in isolation; ω' = fall velocity of discrete particles in suspension consisting of coarse particles at concentrations S_{vc}.

Then we further considered the damping effect caused by flocs on discrete settling particles. When flocs were formed by the connection of clay particles, a great amount of water entrapped in flocs moved together with clay particles as solid. Therefore the effective concentration of flocs, that is, the concentration of clay particles together with entrapped water, is much higher than the concentration of clay particles. The ratio between the volume of clay particles together with entrapped water and the volume of pure clay particles was denoted by α. If the concentration of clay particles was denoted by S_{vf}, then the effective concentration of flocs will be αS_{vf}. The fall velocity of discrete particles in suspension consisting of both coarse particles and clay particles ω'' should be:

$$\omega'' = \omega' (1 - \alpha S_{vf})^{4.65} \tag{5.16}$$

Since the size of flocs was rather smaller (in the order of ten microns), the exponent m is taken as 4.65 here.

Substituting Equation (5.15) into Equation (5.16), one obtained:

$$\omega'' = \omega_0 (1 - S_{vc})^m (1 - \alpha S_{vf})^{4.65} \tag{5.17}$$

B. *Fall velocity of discrete flocs.* Let us consider the settling of flocs in suspension consisting of both discrete flocs and discrete particles. If the concentration of discrete particles is S_{vc} and that of clay particles is S_{vf} the fall velocity of flocs ω_A should be:

$$\omega_A = \omega_{A_0}[1 - (S_{vc} + \alpha S_{vf})]^{4.65}$$

in which $m = 4.65$ is used due to the small size of flocs. ω_{A_0} is the fall velocity of a single floc in isolation. The definition of α is the same as that in Equation (5.16) (Xia & Song, 1983).

This equation can be rewritten as:

$$\omega_A^{1/4.65} = \omega_{A_0}^{1/4.65} \left[1 - \alpha\left(S_{vf} + \frac{S_{vc}}{\alpha}\right) \right] \tag{5.18}$$

Equation (5.18) reveals that there is a linear correlationship between $\omega_A^{1/4.65}$ and $S_{vf} + S_{vc}/\alpha$ although α in it remains an unkonwn value because the volume of flocs could not be measured directly in experiments. But α value, or the volume of flocs could be indirectly deduced as follows.

It can be deduced from Equation (5.18) that when $\omega_A = 0$.

$$1 - \alpha\left(S_{vf} + \frac{S_{vc}}{\alpha}\right)_0 = 0$$

or

$$\alpha = \frac{1}{\left(S_{vf} + \dfrac{S_{vc}}{\alpha}\right)_0}$$

By trial-and-error method a proper α value, with which the experiment dots distribute along a straight line, can be ascertained.

Figure 5.8 was plotted by Xia and Song in this way.

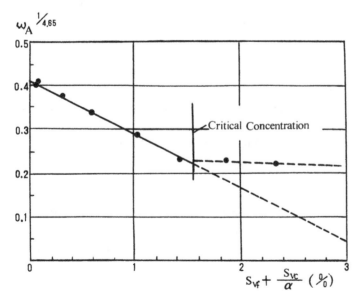

Figure 5.8. Correlationship between settling velocity of flocs and the concentration.

In the region with concentration less than 0.0156 experiment dots fairly distribute along a straight line and Equation (5.18) is proved.

From Figure 5.8 they got

$$\left(S_{vf} + \frac{S_{vc}}{\alpha} \right)_0 = 0.0339$$

Under the condition of their experiments $\alpha = 29.5$. It means that the ratio between the effective concentration of flocs and the clay concentration is 29.5, that is, the volume of entrapped water is 28.5 times the volume of clay particles.

It can also be seen from the figure that $\omega_A^{1/4.65} = \omega_{A_0}^{1/4.65} = 0.41$ as concentration approaches zero. Correspondingly, $\omega_{A_0} = 0.0158$ cm/s, that is, the fall velocity of a single floc in isolation is 0.015 cm/s. It can also been seen that experiment dots start deviating from the straight line as concentration reaches 0.0156 (the corresponding clay concentration is 0.0151 and the total concentration $S_{vc} + S_{vf}$ is 0.03). It means that the flocculent network has been formed and its settling follows another law, which will be discussed later.

5.7 THE SETTLING OF DISCRETE PARTICLES IN FLOCCULENT SUSPENSION

Once the flocculent structure is formed, the damping effect due to the structure is so great that the mutual interference between the discrete coarse particles can be

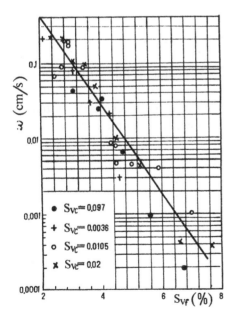

Figure 5.9. Fall velocity of coarse particles in flocculent suspension.

neglected. In Xia & Wang's (1982) experiments although the coarse particle concentration S_{vc} varied by 25 times, no systematic deviation could be detected, as shown in Figure 5.9.

It means that the resistance mainly comes from flocculent structure and the concentration of coarse particles has no significant effect. Qian & Yang (1980), Zhang & Ren (1980), Xu & Wu (1983), Xia & Song (1983) have studied this phenomenon and proposed a number of empirical or theoretical formulae.

In reviewing the history, one will find that at the beginning of this century when people started to study the settling of spheres or other particles in clear water, many empirical or theoretical formulae, such as Stokes' formula, Goldstein's formula and Oseen's formula, were proposed for regions of different Reynolds number. Through several decades practice, all these formulae, except that of Stokes, are seldom in use at present. One relies essentially on the correlation between the drag coefficient C_D and the Reynolds number established by accurate experiments to calculate the settling velocity of a particle. In most cases clay suspension can be properly described as a Bingham fluid. Perhaps it is clearer and simpler to use the same method of attack in dealing with the settling of particles in a Bingham fluid.

Let us start with the settling of a particle in clear water (Newtonian fluid). In laminar flow region the force acting on it can be written as follows:

$$C_D \cdot \gamma_0 \cdot \frac{\pi D^2}{4} \cdot \frac{\omega_0^2}{2g} = 3\pi\mu\omega_0 D$$

It can be deduced from the foregoing equation that:

$$C_D = \frac{24}{Re} \tag{5.19}$$

That is, the drag coefficient C_D is inversely proportional to the reciprocal of the Reynolds number $Re = \omega_0 D/\nu$, in which ν is the kinematic viscosity, $\nu = \mu/(\gamma_0/g)$. In log-log plot Equation (5.19) is a straight line with a slope of $-45°$.

The drag coefficient C_D turns to be a constant in fully developed turbulent region. Between them is a transitional region, in which a smooth curve connects the above mentioned two straight lines, as shown in Figure 5.10.

In Bingham fluid the drag force is caused not only by rigidity η, corresponding to viscosity μ in Newtonian fluid, but also by yield stress τ_B (Wan, 1986).

In laminar flow region the drag force caused by rigidity is assumed to be $3\pi\eta\omega D$, which takes the same form as that in Newtonian fluid. The drag force caused by yield stress τ_B consists of two parts, that is, the resultant vertical force of shear stress τ_B, acting on the surface of the sphere, $(¼)\,\pi^2 D^2\tau_B$ (Qi, 1978); and the resultant vertical force of normal stress, acting on the surface of sphere, which can be deduced by sliding line theory (Ansley & Smith, 1967) as $(⅝)\,\pi^2 D^2\tau_B$. The total drag force in laminar flow region can be written as:

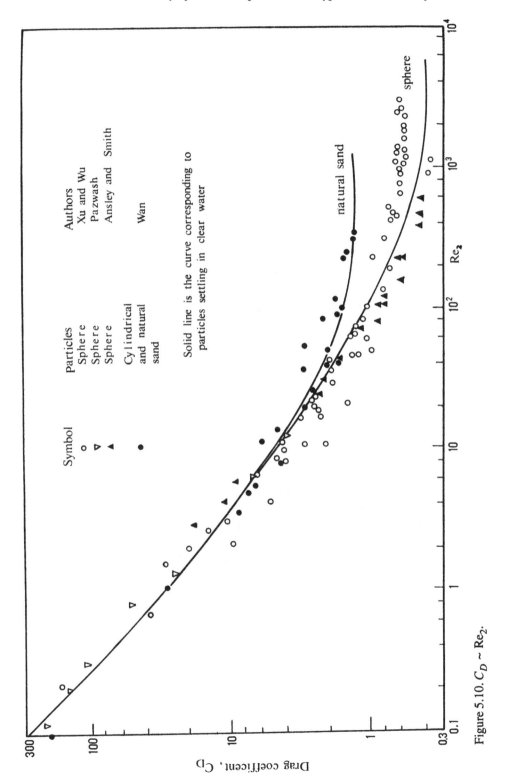

Figure 5.10. $C_D \sim Re_2$.

Table 5.4. The experimental conditions of solid grains settling in Bingham fluid.

Authors	Particles used	Suspensions used
Xu & Wu	Steel balls of 3-12 mm	Clay suspension with specific weight of 1.05-1.35 g/cm^3
Pazwash	Spheres	Slurry
Ansley & Smith	Silver balls	Tomato sauce
Wan	Natural sand of 1.34 mm	Bentonite suspension
	Cylindrical plastic beads:	Kaoline and bentonite
	Length 2.5-3.15 mm	Suspensions
	Diameter 3-3.1 mm	
	Specific gravity 1.29	
	Cylindrical plastic beads:	
	Length 2-2.6 mm	
	Diameter 2.5-2.6 mm	

$$C_D \cdot \gamma_m \frac{\pi D^2}{4} \cdot \frac{\omega^2}{2g} = 3\pi\eta D_\omega + \frac{7}{8}\pi^2 D^2 \tau_B$$

in which γ_m is the specific weight of the clay suspension.

It is assumed that in laminar flow region the drag factor C_D still follows Equation (5.19) but the Reynolds number should be replaced by the following revised Reynolds number Re_2,

$$\mathrm{Re}_2 = \frac{\gamma_m \omega^2}{g\left(\eta\dfrac{\omega}{D} + \dfrac{7\pi}{24}\tau_B\right)} \tag{5.20}$$

In Figure 5.10 available experimental data of the settling spheres or sediment particles in Bingham fluid are plotted in a $C_D - \mathrm{Re}_2$ diagram. The $C_D - \mathrm{Re}_2$ correlationship for particles settling in clear water is plotted in the same diagram. The properties of solid grains and suspensions used in different experiments are listed in Table 5.4. Dots based on experiment data for solid grains settling in suspension distribute just along the curve for clear water with a certain degree of scatter. This implies that the settling of particles in Bingham fluid follows the same rule as that in clear water provided the Reynolds number takes the form of Equation (5.20).

If the concentration of suspension is so high that the submerged weight of a particle is just balanced by the drag force caused by yield stress τ_B, the particle does not settle and it turns to be a neutrally buoyant load. The critical condition is:

$$(\gamma_s - \gamma_f)\frac{\pi D^3}{6} = \frac{7\pi}{8}D^2 \tau_B \tag{2.21}$$

or

$$D = K\frac{6\tau_B}{\gamma_s - \gamma_f}$$

in which $K = 7\pi/8$ (Wan, 1986). But Qian & Yang (1980) and Xu & Wu (1983) gave a K value close to 1 (0.95).

5.8 THE SETTLING OF THE FLOCCULENT STRUCTURE AS AN ENTIRETY – THE CONSOLIDATION PROCESSES OF DEPOSIT

Once all the particles are entrapped by the flocculent structure and a completely homogeneous fluid is formed, the whole flocculent network falls as an entirety with an extremely low velocity. Actually, it is a problem of seepage or consolidation processes, rather than a real problem of sedimentation. That is, water filtrates through the pores of the flocculent network.

Based on such consideration Tsai (1983) studied this problem. The seepage of water through porous medium obeys the famous Darcy's law:

$$\omega = K \frac{\Delta H}{\Delta L} \tag{5.22}$$

In which ΔL is the thickness of the porous medium, ΔH the head difference between BB' and AA', and K is the seepage coefficient and ω the seepage velocity, see Figure 5.11. The piezometric head can be written as:

$$H = Z + \frac{P}{\gamma_0} \tag{5.23}$$

in which Z is elevation, P the pressure and γ_0 the specific unit weight of the clear water.

While the flocculent network is settling in its entirety, the following equation is written down:

$$\Delta H = H_B - H_A = \left(Z_B + \frac{P_B}{\gamma_0}\right) - \left(Z_A + \frac{P_A}{\gamma_0}\right) = \Delta L \cdot \frac{\gamma_s - \gamma_0}{\gamma_0} \cdot S_v$$

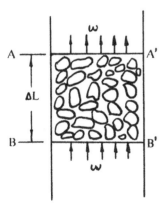

Figure 5.11. The sketch of filtration.

In the deduction the law of static pressure distribution $P_B = P_A + \gamma_m (Z_A - Z_B)$ is used, in which γ_m is the specific weight of muddy water.

Substituting it into Equation (5.22), Tsai got:

$$\omega = K \cdot \frac{\gamma_s - \gamma_0}{\gamma_0} \cdot S_v \qquad (5.24)$$

By dimension analysis he established the following expression of seepage coefficient.

$$K = b\frac{\gamma_0 \alpha^2}{\mu_0} \qquad (5.25)$$

in which μ_0 is the viscosity of water, b the coefficient, and α the length scale of the pores.

The length scale of the pores α is deduced based on a simple model.

It is assumed that the porous flocculent network consists of flocs, which are formed by sediment particles as well as entrapped water. Assume the volume of each floc, $v_2 = \alpha v_0$, in which v_0 is the volume of sediment particles. The number of flocs in unit volume can be calculated from the volumetric concentration S_v.

$$n = \frac{S_v}{v_0}$$

The radius of flocs γ can be calculated from their volume:

$$\gamma = \sqrt[3]{\frac{3}{4\pi} \cdot v_2} = \left(\frac{3\alpha}{4\pi}v_0\right)^{1/3} \qquad (5.26)$$

The average distance between two neighboring particles should be:

$$n^{-1/3} = \frac{v_0^{1/3}}{S_v^{1/3}}$$

And the length scale of pores α should be:

$$\alpha \sim n^{-1/3} - \beta \sqrt[3]{\frac{3\alpha}{4\pi}} \cdot v_0^{1/3}$$

in which β is a coefficient, depending on floc properties, properties of ions in water, temperature, etc.

α can be rewritten as:

$$\alpha \sim v_0^{1/3} \left[\frac{1}{S_v^{1/3}} - \beta \sqrt[3]{\frac{3\alpha}{4\pi}}\right] \qquad (5.27)$$

Substituting Equation (5.27) into Equation (5.25), he obtained:

$$K \sim \frac{\gamma_0}{\mu_0} v_0^{2/3} \left[\frac{1}{S_v^{1/3}} - \beta \sqrt[3]{\frac{3\alpha}{4\pi}}\right]^2$$

and:

$$\omega = b\,\frac{\gamma_0}{\mu_0}v_0^{2/3}\left[\frac{1}{S_v^{1/3}} - \beta \sqrt[3]{\frac{3\alpha}{4\pi}}\right]^2 \left(\frac{\gamma_s - \gamma_0}{\gamma_0}\right)S_v \qquad (5.28)$$

or

$$\omega = b\,\frac{1}{\mu_0}(\gamma_s - \gamma_0)\,v_0^{2/3}\left[1 - \beta \sqrt[3]{\frac{3\alpha}{4\pi}}S_v^{1/3}\right]^2 S_v^{1/3}$$

For a single particle (represented by D_{50}) settling in clear water the fall velocity ω_0 should be:

$$\omega_0 = \frac{1}{18}\cdot\frac{D_{50}^2}{\mu_0}(\gamma_s - \gamma_0) \qquad (5.29)$$

so:

$$\frac{\omega}{\omega_0} = 18b\,\frac{v_0^{2/3}}{D_{50}^2}\left[1 - \beta \sqrt[3]{\frac{3\alpha}{4\pi}}S_v^{1/3}\right]^2 S_v^{1/3}$$

or:

$$\left(\frac{\omega}{\omega_0}S_v^{-1/3}\right)^{1/2} = A - BS_v^{1/3} \qquad (5.30)$$

in which

$$A = (18b)^{1/2}\frac{v_0^{1/3}}{D_{50}^2}$$

$$B = (18b)^{1/2}\beta^3 \sqrt{\frac{3\alpha}{4\pi}}\frac{v_0^{1/3}}{D_{50}^2}$$

Using experiment data obtained at the Institute of Hydraulic Research, Yellow River Conservancy Commission Tsai checked this formula and the agreement is quite all right, see Figure 5.12.

Later Li (1989) did an analysis similar to Tsai's. He obtained Equation (5.24) based on the same consideration. Then he introduced Wang Wenxi's model to express the seepage coefficient K as follows:

$$K = C_s \cdot \frac{\gamma_0 R^2}{\mu} \cdot E \qquad (5.31)$$

where E is void volume, $E = 1 - \alpha S_v$, R radius of capillary channel, C_s shape coefficient, α expanding multiple of granular volume due to the flocculation.

The number of flocs in unit volume is:

$$n = \frac{\alpha S_v}{\dfrac{\pi}{6}\cdot D_j^3}$$

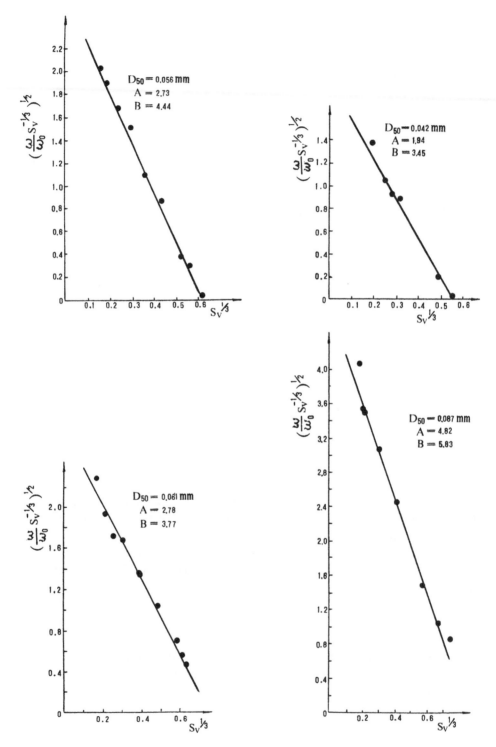

Figure 5.12. The verification of Tsai's theory.

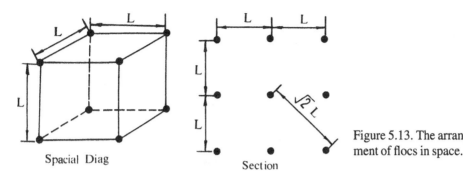

Figure 5.13. The arrangement of flocs in space.

Spacial Diag

Section

in which D_f the diameter of flocs.

Assume that flocs are arranged as a square matrix in space as shown in Figure 5.13.

Then the distance between centers of two neighbouring flocs should be:

$$L = n^{-\frac{1}{3}} = \left(\frac{\pi}{6\alpha}\right)^{\frac{1}{3}} \cdot \frac{D_j}{S_v^{\frac{1}{3}}}$$

and the radius of capillary channel should be:

$$R = a \cdot \frac{L}{2}(\sqrt{2}L - D_f)$$

In which a is coefficient considering the irregular arrangement of flocs.

Substituting all the foregoing formulae into Equation (5.31), he obtained:

$$K = C_s \cdot \frac{\gamma_0}{\mu} \cdot \frac{a^2}{2}\left(\frac{\pi}{6\alpha}\right)^2 D_j^2 \left[1 - \frac{1}{\sqrt{2}}\left(\frac{6\alpha}{\pi}\right)^{\frac{1}{3}} S_v^{\frac{1}{3}}\right]^2 \frac{1}{S_v^{\frac{2}{3}}}(1 - \alpha S_v) \qquad (5.32)$$

Insert it into Equation (5.24), one obtained:

$$\omega = \frac{a^2}{2} \cdot C_s \frac{(\gamma_s - \gamma_0)D_j^2}{\mu}\left(\frac{\pi}{6\alpha}\right)^{\frac{2}{3}}\left[1 - \frac{1}{\sqrt{2}}\left(\frac{6\alpha}{\pi}\right)^{\frac{1}{3}} S_v^{\frac{1}{3}}\right]^2 S_v^{\frac{1}{3}}(1 - \alpha S_v)$$

Let $\frac{\alpha^2}{2} \cdot C_s = A$,

$$\omega = A\frac{(\gamma_s - \gamma_0)D_j^2}{\mu}\left(\frac{\pi}{6\alpha}\right)^{\frac{2}{3}}\left[1 - \frac{1}{\sqrt{2}}\left(\frac{6\alpha}{\pi}\right)^{\frac{1}{3}} S_v^{\frac{1}{3}}\right]^2 S_v^{\frac{1}{3}}(1 - \alpha S_v) \qquad (5.33)$$

In Equation (5.33) there are two unknown coefficients A and α. The author got them experimentally for four kinds of sediment. A general tendency of the variation of A and α has not been obtained yet.

5.9 SOME UNUSUAL SETTLING PHENOMENA ASSOCIATED WITH EXTREMELY HIGH CONCENTRATION OF CLAY PARTICLES

Some unusual settling phenomena associated with extremely high concentration of clay particles have been observed and recorded. In Xia and Wang's experiments the following interesting settling phenomena were recorded.

A. *Uneven settling of uniform coarse particles.* Xia & Wang (1982) recorded obvious uneven settling of uniform coarse particles as shown in Figure 5.14. Sets of experiment runs were conducted. Same nearly uniform coarse particles were used and the concentration of coarse particles S_{vc} kept constant, that is, $S_{vc} = 0.36\%$. But the clay concentration changed in different runs. The distribution of fall velocities of coarse particles were measured and plotted in Figure 5.14. Uniform coarse particles settle evenly at low clay concentration, for instance, under the condition of $S_{vf} = 0.293\%$. Under the condition of higher clay concentration fall velocities of coarse particles reduced and the distribution of fall velocities of coarse particles became more and more uneven. When the clay concentration was very high, the settling of uniform coarse particles turned to be even again.

The explanation given by the authors are as follows:

It was seen under microscope that the flocculent structure was uneven. It was denser at some places and looser at other places. Settling of coarse particles in flocculent network is a stochastic phenomenon. Some particles settled smoothly through looser flocculent structure with vertical channels and hence they had higher fall velocity. Some particles settled through denser flocculent structure

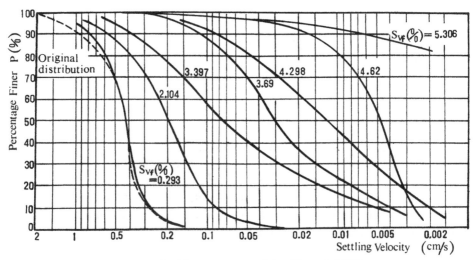

Figure 5.14. Uneven settling of uniform coarse particles ($S_{vc} = 0.0036$).

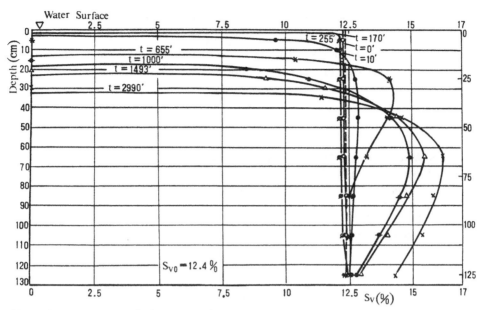

Figure 5.15. Reverse vertical concentration profiles.

with zigzag channels and therefore they had much lower fall velocity. And some particles were supported by flocculent structure and settled together with the flocculent structure with extremely low velocity. Although these coarse particles were uniform in size, their settling behaviour could be rather different. Furthermore, the afore mentioned picture was not a static one. With the breaking of existing links of the flocculent structure and the formation of new links, coarse particles changed their settling behavior correspondingly.

B. *Reverse concentration profiles*. It is not accidental that reverse concentration profiles have been recorded for many times. That is, the concentration was higher in the upper layers but lower in the lower layers. An example is shown in Figure 5.15 (Xia & Song, 1983). In that run of experiment the original concentration S_{vo} was 12.4%, concentration profile measured at different moment (minute) are shown in the figure.

Starting with uniform concentration distribution at the very beginning, reverse concentration profile appeared at 255 minutes and clearly occurred at 655 minutes. It maintained until the end of the settling.

The authors consider that this anomalous phenomenon is closely related to the non-uniformity of the flocculent structure and the existence of discrete coarse particles. Due to the non-uniformity of the flocculent structure the amount of discrete particles located at different layers of the flocculent structure is different. The seepage is weaker and the seepage velocity is smaller where more discrete particles are concentrated. In other words, the fall velocity of particles is smaller there. On the other hand, the seepage is stronger and the fall velocity is larger where less discrete particles exist. As a result, more discrete particles will be amassed into layers with higher concentration.

REFERENCES

Ansley, R.W. & T.N. Smith 1967. Motion of spherical particles in a Bingham plastic. *J. of AICHE*, Vol. 13(6): 1193-1196.

Guo, M., Y. Zhuang 1963. Fluidization-motion of uniform spheres and fluid in vertical system (in Chinese). *Science Press*, p. 1963.

Hawksley, P.G.W. 1951. The effect of concentration on the settling of suspensions and flow through porous media. In *Some Aspects of Fluid Flow, Edward Arnold and Co., London*, pp. 114-135.

Li, Y. 1989. Capillary-seepage model related to the interface settling of fine sediments. *Proc. of 4th International Symposium on River Sedimentation*, Vol. 1: 555-562.

Moliboxino, 1956. The viscosity of suspensions. *Chemistry Engineering*, Vol. 20(2) (in Japanese).

Qian, N. (Nien Chien) 1980. Preliminary study on the mechanism of hyperconcentrated flow in north-west region of China (in Chinese). *Selected papers of the symposium on sediment problems on the Yellow River*, Vol. 4: 244-267.

Qian, N. (Ning Chien), Z. Wan, Y. Qian 1979. The flow with heavy sediment concentration in the Yellow River Basin (in Chinese). *Journal of Qinghua University*, Vol. 19(2): 1-17.

Qian, N. (Ning Chien) & Z. Wan 1986. A critical review of the research on the hyperconcentrated flow in China. *Series of publication IRTCES*.

Qian, Y., W. Yang et al. 1980. Basic characteristics of flow with hyperconcentration of sediment (in Chinese). *Proc. of the International Symposium on River Sedimentation*, pp. 175-184.

Qi, P. 1978. The mechanism of the effect of very fine sediment particles on the sediment-carrying capacity (in Chinese). *Selection of the Institute of Hydraulic Research, Yellow River Conservancy Commission*, 1978, No. 2.

Richardson, J.F. & W.N. Zaki 1954. Sedimentation and fluidization Pt. 1. *Trans Inst. Chem. Engrs.*, Vol. 32: 35-53.

Sa, Y. 1965. Preliminary mechanics of sediment motion (in Chinese). *Industry Press*, pp. 302.

Shaanxi Provincial Institute of Hydraulic Research 1977. Experiments on the settling of hyperconcentrated non-uniform sediment (in Chinese). *Research Report of SPIHR*.

Steinour, H.H. 1944. Rate of sedimentation: (1) Nonflocculated suspensions of uniform spheres; (2) Suspensions of uniform size angular particles; (3) Concentrated flocculated suspensions of powders. *Industrial and Engin. Chem.*, Vol. 36(7, 9 and 10): 618-624, 840-847, 901-907.

Tsai S. 1983. The velocity of the collective motion of sedimentation of sand and clay (in Chinese). *Applied Mathematics and Mechanics*, Vol. 4(3): 341-346.

Wan, Z. 1986. Fall velocity of coarse particles in clay slurry (in Chinese). *Journal of Sediment Research*, No. 1: 47-55.

Wan, Z. & S. Sheng 1978. Phenomena of hyperconcentrated flow on the stem and tributaries of the Yellow River (in Chinese). *Selected papers of the symposium on sediment problems on the Yellow River*, Vol. 1: 141-158.

Wang, Z. 1987. Buoyancy force in solid-liquid mixture. *Proc. of Technical Session A, 22th Congress, IAHR*, pp. 86-91.

Xia Z. & G. Song 1983. Settling properties of sediments composed of cohesive and non-cohesive particles (in Chinese). *Proc. of 2nd International Symposium on River Sedimentation*, pp. 253-264.

Xia, Z. & G. Wang 1982. The settling of non-cohesive particles in a flocculated suspension (in Chinese). *Journal of Sediment Research*, No. 1: 14-23.

Xu, M. & D. Wu 1983. The analysis of settling characteristics of a spherical particle in Bingham fluid (in Chinese). *Journal of Hydraulic Engineering*, No. 11: 29-36.

Zhang, H., Z. Ren et al. 1980. Settling of sediment and the resistance to flow at hyperconcentrations (in Chinese). *Proc. of the International Symposium on River Sedimentation*, pp. 185-192.

CHAPTER 6

Hyperconcentrated pseudo-one-phase flow

6.1 VELOCITY PROFILES

6.1.1 *Laminar flow*

A hyperconcentrated flow is non-Newtonian if the sediment consists of a certain amount of clay particles. As discussed in Chapter 4, the rheologic properties of the hyperconcentrated flow approximately follows the Bingham model:

$$\tau = \tau_B + \eta \frac{du}{dy} \tag{6.1}$$

If the yield stress τ_B is so large that all particles in the flow are smaller than D_o,

$$D_o = k \frac{\tau_B}{\gamma_s - \gamma_f} \tag{6.2}$$

the sediment in the flow is transported as neutrally buoyant load and the flow can be treated as one-phase flow. In such case the concept of 'sediment-carrying capacity' is meaningless. The amount of sediment transported by the flow depends only on the incoming sediment amount and boundary resistance. Dynamic characteristics of the flow depends on the rheological properties of the mixture, which in turn depend on the content of clay and silt, and cohesionless coarse particles have little influence on the flow.

Velocity distribution of laminar flow of the sediment suspension is much different from that of Newtonian fluid because of the yield stress. In the upper central part of an open channel flow where the shear stress is less than the yield stress, there is no velocity gradient. The mixture there moves entirely at an uniform velocity u_p and forms a flow plug.

Considering a Bingham fluid flowing in an open channel with slope J, one can establish the following equation if the wall resistance is negligible,

$$\gamma_m(H - y)J = \tau_B + \eta \frac{du}{dy} \tag{6.3}$$

92

where γ_m is the specific weight of the mixture, H the depth and y the distance from the bed. By integration with the boundary condition: $u = 0$ at $y = 0$, a theoretical velocity profile is obtained as follows:

$$u = \frac{y}{2\eta}(2\gamma_m HJ - \gamma_m yJ - 2\tau_B), 0 \le y \le H - \frac{\tau_B}{\gamma_m J} \tag{6.4}$$

Equation (6.4) can be rewritten as

$$\frac{u_p - u}{u_p} = \left(1 - \frac{\gamma_m yJ}{\gamma_m HJ - \tau_B}\right)^2, 0 \le y \le H - \frac{\tau_B}{\gamma_m J} \tag{6.5}$$

where u_p is the maximum velocity in the profile and equal to the velocity of the plug zone. In the plug zone, $y > H - \tau_B / \gamma_m J$, the fluid moves at the velocity u_p given by,

$$u = u_p = \frac{\gamma_m J}{2\eta}\left(H - \frac{\tau_B}{\gamma_m J}\right)^2, H - \frac{\tau_B}{\gamma_m J} < y \le H \tag{6.6}$$

Equations (6.5) and (6.6) are the velocity profile of two-dimensional open channel flow. On the other hand, if the width/depth ratio of the flow is small and the plug zone is relatively great, the transverse velocity distribution on the channel surface can be deduced. A schematic diagram is shown in Figure 6.1. Take a thin layer near the surface with a thickness, a. The resistance shear stress on the bottom plane of the layer τ_a is negligible because the bed resistance reduces linearly with distance from the bed. The shear stress equals τ_B at $y = y_o$ and equals zero at $y = H$. Considering a volume of the fluid with a length, L, along the channel, a width, $2z$, in central part, and a thickness, a, near the surface, one can establish the following equation:

$$2azL\gamma_m J = 2aL\left(\tau_B - \eta \frac{du}{dz}\right) \tag{6.7}$$

Integrating the equation with the boundary conditions:

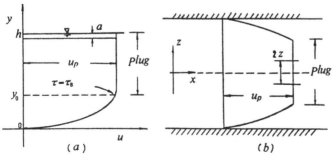

Figure 6.1. Schematic diagram of transverse distribution of surface velocity.

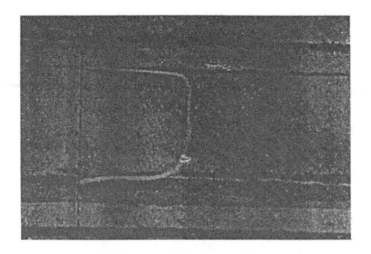

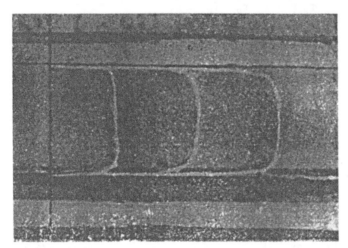

Figure 6.2. Surface velocity distribution of mud flow.

$u = 0$ for $z = B/2$;

$u = u_p$ for $z \leq \tau_B / \gamma_m J$ $\qquad$ (6.8)

one obtains the following transverse velocity distribution on the channel surface:

$$\frac{u_p - u}{u_p} = \left[\frac{\gamma_m J z - \tau_B}{\gamma_m J \dfrac{B}{2} - \tau_B} \right]^2, \quad \frac{B}{2} \geq z \geq \frac{\tau_B}{\gamma_m J} \qquad (6.9)$$

$$u = u_p = \frac{\left(\tau_B - \gamma_m J \dfrac{B}{2} \right)^2}{2 \eta \gamma_m J}, \quad 0 \leq z < \frac{\tau_B}{\gamma_m J} \qquad (6.10)$$

The transverse velocity distribution on the surface is similar to the vertical distribution. However, visualization of the vertical distribution is difficult because of opacity of the non-Newtonian fluid, but the transverse velocity distribution on the surface can be easily visualized.

Figure 6.2 shows three photos presented by Hisinko (1951). These photos were taken as hyperconcentrated clay suspension flowed through a flume. The first two photos show the surface velocity distributions at low flow rate. Among them the second one was exposed three times with time interval 3 seconds. It is clearly shown that the central part was a plug and moved at an uniform velocity. The third photo shows the velocity distribution at high flow rate. The flow was in transitional region and no plug zone was observed.

Figure 6.3 shows surface velocity distribution of hyperconcentrated flow in the Luohui Irrigation Canal, Northwest China, through which hyperconcentrated flood was diverted from the North Luohe River for irrigation (Xu & Wan, 1985).

Figure 6.3. Surface velocity distribution of hyperconcentrated flow in the Luohui Irrigation Canal.

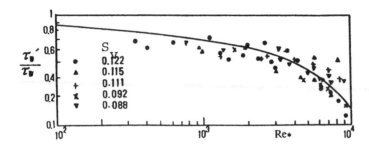

Figure 6.4. Variation of the yield stress in turbulent flow.

About 80% of the width was plug and velocity gradient existed only in zones close to the banks.

It seems from the third photo in Figure 6.2 that there was no plug if the flow rate was high, possibly because the yield stress reduced at high flow rate. Wang & Fei (1989) estimated the yield stress by measuring the width of the plug zone on the surface and found that the yield stress decreased with increasing Reynolds number Re* and vanished as Re* was larger than 30 000. Figure 6.4 shows the measured results, in which τ_B' and Re* were calculated with the following formulas:

$$\tau_B' = \frac{\gamma_m JHZ}{H + \dfrac{B}{2}}$$ (6.11)

$$Re^* = \frac{4\rho_m RU}{\eta}\left[1 - \frac{3\,\tau_B}{2\,\tau_o} + \frac{1}{2}\left(\frac{\tau_B}{\tau_o}\right)^2\right]$$ (6.12)

where τ_B' is the yield stress determined from the width of the plug and τ_B is the yield stress from rheologic measurement; R is hydraulic radius, U average velocity, τ_o shear stress on the boundary and Z half of the plug width. One can see from the figure that the ratio τ_B'/τ_B decreases with increasing Reynolds number. It reduces to 0.15 at Re* = 10^4. The reduction of the yield stress at high Reynolds number was attributed to the destruction of flocculent structure at high flow velocity.

The yield stress of clay mud is not constant. It reduces in fully-developed turbulent flow, and on the other hand, it may increase at some circumstances. Both in rivers and flumes a phenomenon so called 'stagnant layer' has been observed. Figure 6.5a shows a chair form velocity profile measured in a flume experiment with sediment suspension from the Yellow River (Ren, 1985). The upper part was plug and there was no velocity gradient. Under the plug was a shear layer where the velocity gradient was quite high. A sluggish layer developed near the bed where both velocity and velocity gradient were very small. The sluggish layer may further develop into stagnant layer at some circumstances. Figure 6.5b shows

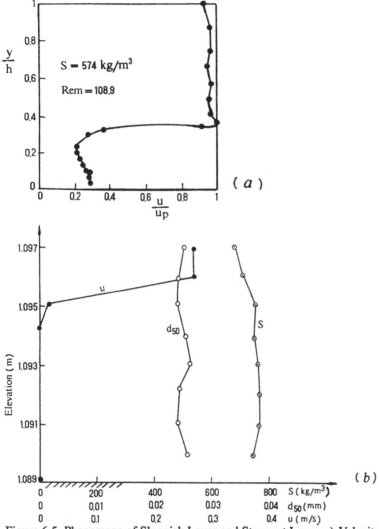

Figure 6.5. Phenomena of Sluggish Layer and Stagnant Layer. a) Velocity distribution with sluggish layer in laboratory experiment, b) Stagnant layer observed in the Bajiazui Reservoir.

the velocity profile, sediment concentration distribution and median diameter distribution measured in the Bajiazui section in a tributary of the Yellow River (Fang & Hu, 1984). The lower mud layer came to standstill while the upper layer remained flowing. Sediment concentration and median diameter in the stagnant layer were essentially the same as that in the upper layer. Such an surprising velocity profile was not understandable without being aware of change in rheologic properties of the sediment suspension during flowing.

A phenomenon so called 'rheopexy' or 'antithixotropy' was observed in a

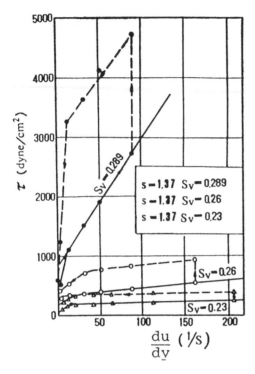

Figure 6.6. Rheopexy phenomenon in hyperconcentrated clay suspensions.

study of rheology of clay suspension (Wang & Qian, 1984). Figure 6.6 shows measured flow rheogram of clay suspensions at high concentrations ($S_v = 0.23$-0.29). The clay suspensions behaved at first like ordinary Bingham fluid possessing a certain yield stress and certain rigidity coefficient. As the shear rate was over a critical value, however, the shear stress at a certain shear rate did not remain constant. It increased from a certain value to a value nearly doubled. This phenomenon is defined as 'rheopectic' although it does not adhere strictly to the definition. The flow curve after rheopexy had the same slope with before (Figure 6.6), or the rigidity coefficient did not change and the increase in shear stress was referred to increase in the yield stress.

The phenomenon of stagnant layer and the chair form velocity profile can be interpreted with the change in the yield stress due to rheopexy. Assuming that the original velocity distribution is normal as shown in Figure 6.7, and the critical shear rate for rheopexy is D_k. A layer of mud under elevation y_o gains a greater yield stress when rheopexy occurs. The yield stress of the mud in the upper part remains the original value because the shear rate there is smaller than the critical value D_k. The following three inequalities may hold:

$$\tau_{B1} > \gamma_m HJ$$
$$\tau_B < \gamma_m (H - y_o) J$$
$$\tau_B > \gamma_m (H - y_1) J \tag{6.13}$$

where τ_{B1} is the yield stress of the lower mud layer after rheopexy and τ_B is the original value. The first inequality interprets that mud in the lower layer $(y < y_o)$ does not flow because the tractive shear stress is less than the yield stress τ_{B1}. The second inequality means that mud between $y = y_o$ and $y = y_1$ is subjected to shear and there exists velocity gradient. And the third inequality indicates that there is no shear rate in the upper layer $(y > y_1)$ and the upper layer moves at an uniform velocity.

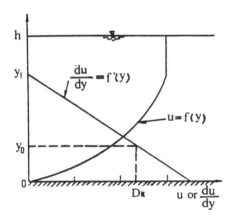

Figure 6.7. Velocity and velocity gradient distributions in Bingham open channel flow.

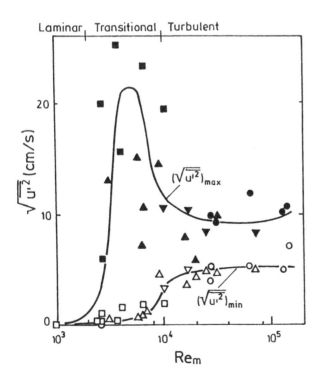

Figure 6.8. Regimes of hyper-concentrated flow.

6.1.2 *Turbulent flow*

A hyperconcentrated pseudo-one-phase flow also transforms into turbulent flow
if Reynolds number is large. Wang et al. (1992) found from experiments with
hyperconcentrated clay suspensions in an open channel that the flow began to
develop into turbulence at $Re_m = 2000$ and developed into fully turbulent one if
$Re_m > 10\ 000$. The flow was in transitional region if $Re_m = 2000$-$10\ 000$. It was
turbulent in a zone near the boundary and remained laminar or intermittent
turbulent in the upper zone. Figure 6.8 shows the maximum and minimum root
mean square of fluctuating velocity as a functions of Re_m that is defined by

$$Re_m = \frac{4\rho_m H U}{\eta\left(1 + \dfrac{1}{2}\ \dfrac{\tau_B H}{\eta U}\right)} \tag{6.14}$$

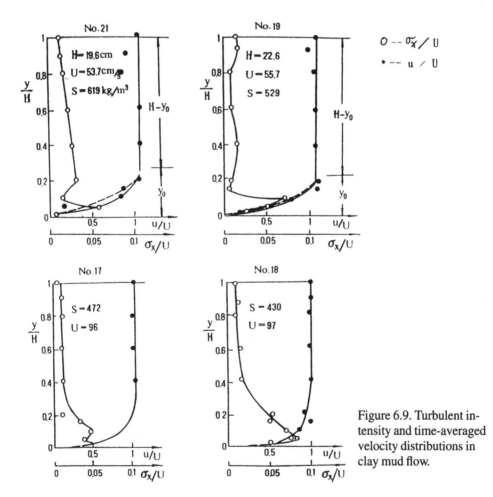

Figure 6.9. Turbulent in-
tensity and time-averaged
velocity distributions in
clay mud flow.

One can see from the figure that in the transitional zone the maximum turbulence was very strong but the minimum turbulence nearly zero. The flow was partly turbulent and partly laminar.

Figure 6.9 shows velocity profiles and distributions of turbulence intensity of hyperconcentrated flows measured by Wang et al.(1983), where σ_x is the root mean square of fluctuating velocity in the flow direction. There still existed a plug in the upper zone where there was no velocity gradient and the turbulence intensity was nearly zero. In the lower layer close to the bed there was velocity gradient and the flow was turbulent. The maximum turbulence intensity occurred at $y=0.05$-$0.15\,H$. The results also indicated that the turbulence intensity increased with increasing Re_m and decreased with increasing sediment concentration.

In the fully developed turbulent flow the mean velocity distribution seemed to still follow the logarithmic formula but the velocity gradient or the constant κ is different from that of clear water flow. Figure 6.10 shows a few velocity profiles in turbulent region measured in a flume experiment (Yang & Zhao, 1983). Points followed logarithmic lines in the semi-log plot. Table 6.1 gives main parameters of the experiment. A few velocity profiles in the transitional region are also shown in the figure, there were still plug zones in these cases. Under the plug zone the flow was turbulent and the velocity profiles were logarithmic. In Table 6.1, H_o is the thickness of the plug zone.

Figure 6.11 shows velocity profiles of fully-developed turbulent hyperconcentrated flows in the Wudin River (Qian & Wan, 1983). The constant κ varied with

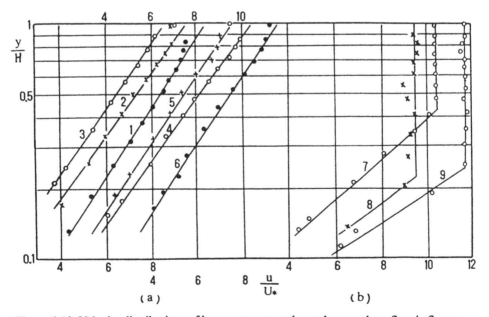

Figure 6.10. Velocity distributions of hyperconcentrated pseudo-one-phase flow in flume.

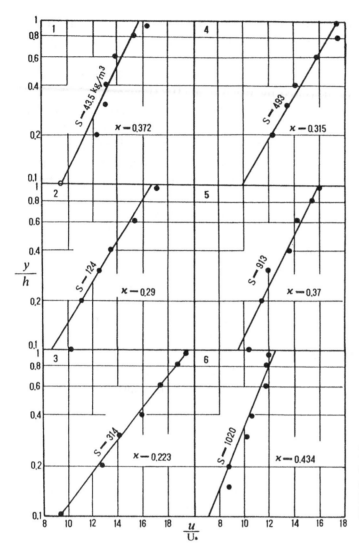

Figure 6.11. Velocity distributions of fully-developed turbulent hyperconcentrated flow in the Wudin River.

sediment concentration as shown in Figure 6.12. κ reduced from 0.4 for clear water flow to a minimum value about 0.25 for sediment suspension of concentration about 300 kg/m^3.

For fully-developed turbulent hyperconcentrated flow in pipes or closed channels, the velocity profile also follows the logarithmic formula. Figure 6.13 shows velocity profiles of turbulent and laminar pseudo-one-phase flows in a closed conduit with a cross section 0.18 × 0.1 m^2 (Song et al., 1986). The sediment used in the experiment was from the Yellow River. The median diameter was 0.0045 mm. The constant κ of the velocity profiles in the turbulent flow was between 0.2 and 0.3. The thickness of the plug in the pipe flow is given by:

Table 6.1. Experiment data of homogeneous hyperconcentrated flow in flume (Yang & Zhao, 1983)

No.	U (m/s)	H (m)	Slope J	S (kg/ m³)	H_o/H	κ	Remarks
1	1.397	0.1432	1/50	515	0	0.335	
2	1.163	0.1096	1/50	315	0	0.311	
3	0.905	0.0798	1/50	470	0	0.330	Fully-developed turbulent
4	1.245	0.1205	1/50	590	0	0.297	
5	1.016	0.0944	1/50	550	0	0.320	
6	0.749	0.1095	1/100	440	0	0.331	
7	1.075	0.1395	1/100	590	0.55	0.197	
8	0.745	0.1395	1/200	550	0.75	0.151	Transitional
9	1.002	0.1635	1/20	550	0.75	0.141	

Note: The sediment used in the experiment was from the Yellow River with D_{50} = 0.034 mm and D_{28} = 0.01 mm.

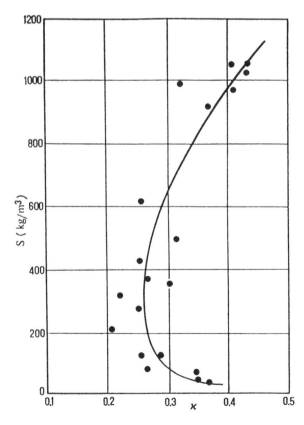

Figure 6.12. Variation of von Karman constant with sediment concentration.

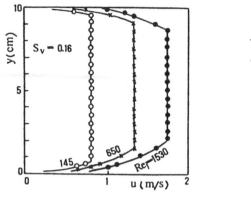
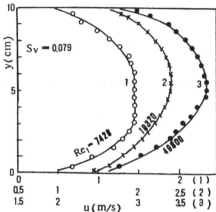

Figure 6.13. Velocity distributions of turbulent and laminar flows in pipe-line.

$$b = \frac{\tau_B}{\tau_o}H = \frac{\tau_B H}{\gamma_m R J_m} \tag{6.15}$$

where τ_o is the shear stress on the pipe wall, H (= 10 cm) is the distance from the bottom wall to the top wall, R the hydraulic radius and J_m the energy slope. The measured data in laminar flows conformed to the equation but the measured b in turbulent flows was smaller because of reduction in yield stress owing to turbulence.

6.2 RESISTANCE

6.2.1 *Laminar flow*

In two dimensional flow, integration of Equation (6.4) and (6.6) yields flow discharge per width q,

$$q = \frac{\tau_o H}{3\eta}\left[1 - \frac{3}{2}\frac{\tau_B}{\tau_o} + \frac{1}{2}\left(\frac{\tau_B}{\tau_o}\right)^3\right] \tag{6.16}$$

where

$$\tau_o = \gamma_m H J \tag{6.17}$$

is the shear stress on the bed. Because a Bingham fluid can flow only as τ_o is larger than τ_B, the third term in the brackets is often much smaller than 1 and can be neglected. Therefore, Equation (6.16) can be written as:

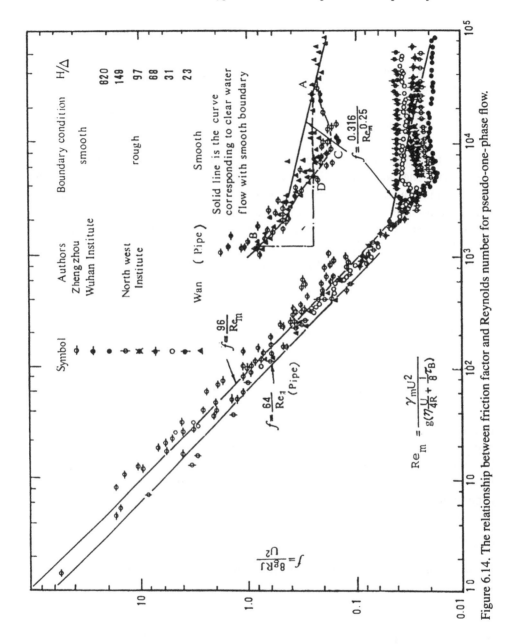

Figure 6.14. The relationship between friction factor and Reynolds number for pseudo-one-phase flow.

$$f = \frac{96}{\mathrm{Re}_m} \tag{6.18}$$

in which the Reynolds number has been given by Equation (6.14), f is the friction factor defined by:

$$f = 8\frac{gHJ}{U^2} = 8\frac{U^{*2}}{U^2} \tag{6.19}$$

Figure 6.14 shows measured friction factor as a function of Re_m from different researchers. The data were collected from flume and pipe experiments. Since the flows in flume were not two dimensional, the depth H in Equations (6.19) and (6.14) was replaced by hydraulic radius R. It turns out that the $f - \mathrm{Re}_m$ correlation of hyperconcentrated flow is the same as the $f - \mathrm{Re}$ correlation in Newtonian fluid flow (Qian & Wan, 1985).

For pipe flow the friction factor-Reynolds number correlation is given by:

$$f = \frac{64}{\mathrm{Re}_1} \tag{6.20}$$

where the Reynolds number has the following form:

$$\mathrm{Re}_1 = \frac{4\rho_m UR}{\eta\left(1 + \frac{2}{3}\frac{\tau_B R}{\eta U}\right)} \tag{6.21}$$

It can be seen from Figure 6.14 that the measured points of laminar flow in flumes and pipes follow well the two formulas (6.18) and (6.20). The results indicate that the resistance of laminar hyperconcentrated flow obeys the same rule as a Newtonian fluid if the yield stress and the rigidity coefficient are taken into account in the Reynolds number.

6.2.2 Turbulent flow

In turbulent hyperconcentrated flow over smooth boundary the friction factor follows Blasius formula:

$$f = 0.316\,\mathrm{Re}_m^{-\frac{1}{4}} \tag{6.22}$$

Nevertheless, if the Reynolds number is larger than 8000-10 000, the friction factor is independent of Re_m but only related to boundary roughness. Since the effective viscosity of hyperconcentrated flow is high, many roughness elements are too small compared with the viscous sublayer. Yang & Zhao (1983) glued cubic concrete blocks of $20\times20\times20$ mm^3 on a flume bed and conducted an experimental study on the resistance of hyperconcentrated flow. Figure 6.15 presents the measured friction factor as a function of relative roughness R/k_s in which k_s is the size of the roughness elements. Some dots of clear water flow are

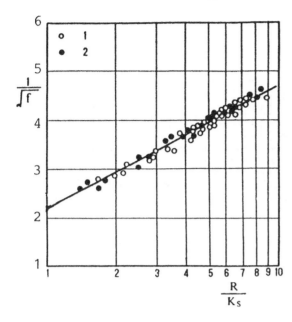

Figure 6.15. Friction factor versus relative roughness: 1. Hyperconcentrated flow; 2. Clear water.

also plotted in the figure as a comparison. The results illustrates that in fully developed turbulent flow the friction factor-relative roughness correlationship of hyperconcentrated flow is the same as that of clear water flow.

In turbulent water flow on rough boundary the logarithmic velocity formula

$$\frac{U}{U^*} = 5.75 \log \left(12.27 \frac{H}{k_s} \right) \tag{6.23}$$

yields the following friction factor-relative roughness relation

$$\sqrt{\frac{8}{f}} = 5.75 \log \left[12.27 \frac{H}{k_s} \right] \tag{6.24}$$

which can be rewritten as:

$$\sqrt{\frac{1}{f}} = 2.03 \log \left(\frac{H}{k_s} \right) + 2.2 \tag{6.25}$$

As shown in Figures 6.15 and 6.16, the measured data of hyperconcentrated flow agree with this relation quite well. It suggests that the yield stress and the rigidity coefficient have no influence on the friction factor as long as the flow is fully turbulent.

It must be noted that because the size of flume and pipe was limited and the velocity in the experiments was also confined by the recirculation system, it was seldom for hyperconcentrated flow to develop into a fully turbulent one. The data corresponding to fully developed turbulence in Figure 6.16 were in fact collected

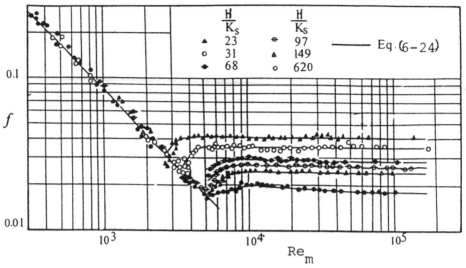

Figure 6.16. Friction factor versus Re and Relative smoothness H/K_s.

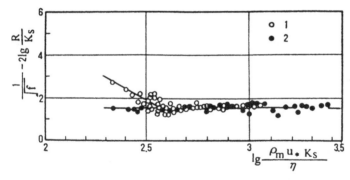

Figure 6.17. Friction factor versus relative roughness and roughness Reynolds number: 1. hyperconcentred pseudo-one-phese flow; 2. Clear water.

in flows with sediment concentration less than 270 kg/m^3 and of small yield stress. It remains to be further studied for the resistance of flow of sediment suspension with high rigidity coefficient and high yield stress at high Reynolds number.

In the transitional zone the friction factor depends both on the Reynolds number and the relative roughness. For flow with smooth boundary measured points follow the Blasius formula. For flow with rough boundary the friction factor is a function of Re$_m$ and R/k_s. Yang & Zhao (1983) suggested the following formula based on their experimental results:

$$\frac{1}{\sqrt{f}} - 2\log\left(\frac{R}{k_s}\right) = 13.81 - 4.71\log\left(\frac{\rho_m U^* k_s}{\eta}\right) \tag{6.26}$$

where R is the hydraulic radius referring to bed friction, $\rho_m U^* K_s/\eta$ is called roughness Reynolds number and denoted by Re*. The measured data of hyper-concentrated flow and clear water flow are shown in Figure 6.17. In the transitional zone, log Re* = 2.3 − 2.6, the measured data in hyperconcentrated flow follow Equation (6.26) but the data of clear water flow do not.

6.2.3 *Other approaches*

Quite a few researchers adopt Hedstrom number He to characterize the relative importance of the yield stress. He is defined by (Hedstrom, 1952):

$$He = \frac{\rho_m \tau_B (4H)^2}{\eta^2} \tag{6.27}$$

With the definition of Hedstrom number Equation (6.16) can be reshuffled into

$$\frac{1}{Re} = \frac{f}{96} - \frac{He}{8\,Re^2} + \frac{8}{3}\left(\frac{He}{Re^2}\right)^3 \frac{1}{f^2} \tag{6.28}$$

where Re is different from Re_m and is defined by

$$Re = \frac{4\rho_m UH}{\eta} \tag{6.29}$$

Equation (6.28) presents a correlationship between the friction factor f and Reynolds number and Hedstrom number for laminar flow of Bingham fluid. Data obtained in a flume, as shown in Figure 6.18, roughly agree with the equation (Wan et al., 1979).

For laminar flow of Bingham fluid in pipes, the following equation can be established:

$$\frac{1}{Re} = \frac{f}{64} - \frac{1}{6}\frac{He}{Re^2} + \frac{64}{3}\left(\frac{He}{Re^2}\right)^4 \frac{1}{f^3} \tag{6.30}$$

where

$$Re = \frac{2\rho_m UR_p}{\eta}; \quad He = \frac{2\tau_B \rho_m R_p^2}{\eta^2};$$

$$f = \frac{4R_p J_m}{\rho_m U^2} \tag{6.31}$$

where R_p is the radius of the pipe and J_m is the pressure gradient. Hua (1988) solved f from the equation:

$$f = \frac{1}{2}\left[\frac{1}{4}(b - \sqrt{8Y + b})^2 - 4Y + \frac{4bY}{\sqrt{8Y + b}}\right]^{1/2} - \frac{1}{4}(b - \sqrt{8Y + b})^2 \tag{6.32}$$

in which Y is given by:

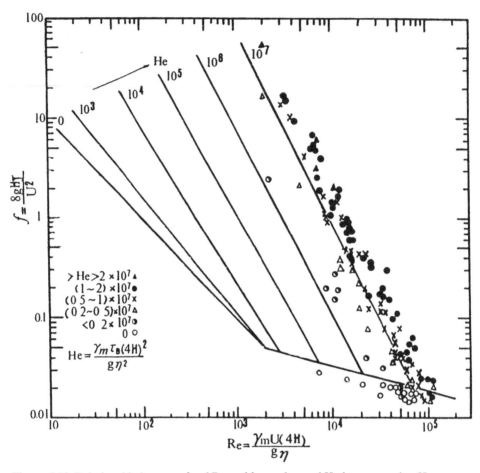

Figure 6.18. Relationship between f and Reynolds number and Hedstrom number He.

$$Y = \left[\frac{cb^2}{16} + \sqrt{\left(\frac{cb^2}{16}\right)^2 - \left(\frac{c}{3}\right)^3}\right]^{1/3} + \left[\frac{cb^2}{16} - \sqrt{\left(\frac{cb^2}{16}\right)^2 - \left(\frac{c}{3}\right)^3}\right]^{1/3} \tag{6.33}$$

and b and c are given by

$$b = -64\left(\frac{1}{6}\frac{He}{Re^2} + \frac{1}{Re}\right), c = 64\frac{2}{3}\frac{He^4}{Re^8} \tag{6.34}$$

The theoretical solution is much more complicated than Equation (6.20) and gives smaller value of f. The larger the Hedstrom number, the greater the difference between Equations (6.32) and (6.20). For He/Re = 0.5, the f value given by Equation (6.20) is about 25-30% greater than the value given by Equation (6.32).

6.2.4 *Drag reduction*

By adding minute quantities of long chain macromolecules like polyacrylamide (PAA) or polystyrene (PS) to a turbulent pipe flow of e.g. pure water or toluene the pressure loss along the pipe is reduced when the flow rate is kept constant or the flow rate is increased when the pressure loss is kept constant. This effect of frictional drag reduction by polymer additives in turbulent flow was first observed by Toms (Toms, 1948) and therefore is often called Toms phenomenon. Enlightened by the fact of drag reduction owing to long chain structure of polymer molecules, some researchers studied possible drag reduction by adding fine sediment in hyperconcentrated flow. It is thought that the flocculent structure of clay suspension might play a similar role as polymer does. A few of reports about the drag reduction have been published. Hou & Yang (1983) designed an experiment with a rotating disk in clay suspension to study the drag reduction by adding fine sediment. They found that the friction factor of the disk rotating in clay suspension is larger in laminar flow but smaller in turbulent flow than that in clear water. Figure 6.19 presents the experimental results, in which the friction factor λ and Reynolds number Re are defined by the following formulas:

$$\lambda = \frac{4M}{\rho_m \omega^2 r^5} \tag{6.35}$$

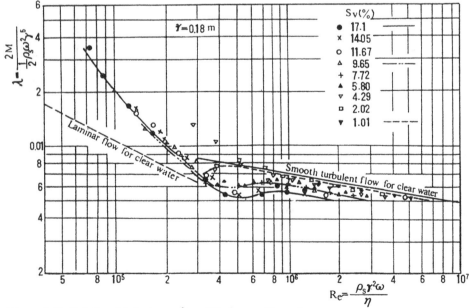

Figure 6.19. Relationship between λ and Re for muddy water.

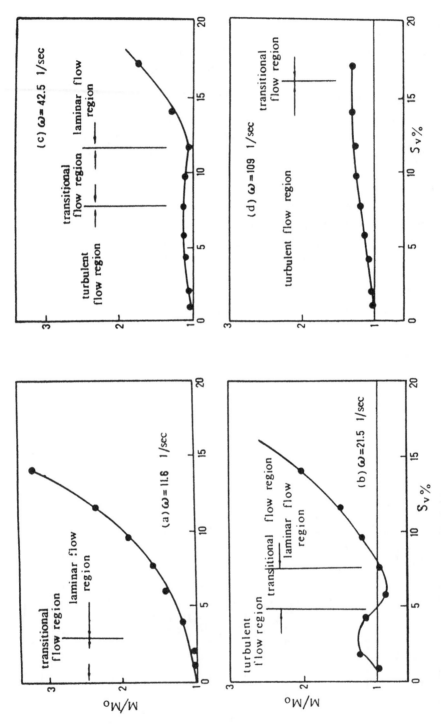

Figure 6.20. The comparison of the moment required for keeping a disk rotating in clear water and in slurry.

$$\text{Re} = \frac{\rho_m r^2 \omega}{\eta} \qquad (6.36)$$

where M is moment of resistance, ω angular velocity of the rotating disk and r the radius of the disk.

It can be seen from the figure that the friction factor in clay suspension was apparently smaller than that in clear water if the concentration was higher than 5% and the flow was in transitional region. The friction factor was also smaller in turbulent region but not so much. It is to be noted that, however, the smaller friction factor in clay suspension does not imply less energy consumption. The comparison of friction factor at the same Reynolds number is not reasonable. Direct comparison of resistance or energy consumption at the same rotating speed is of practical significance. Figure 6.20 presents a direct comparison of the moments of resistance for rotation of the disk in clay suspension and clear water, where M/M_o is the ratio of the moment in clay suspension to that in clear water at the same speed. Most measured points were above the line $M/M_o = 1$. Only one point in the transitional region was bellow the line.

Polymer molecules can be stretched by high strain in the region near the wall. After initial stretching the molecules can interact with turbulence and reduce turbulence intensity. When concentration of polymer molecules is above 0.05 ppm, the molecules connect with each other and produce a network. As a result, some eddies that exist in clear water flow are suppressed, so that the turbulent shear stress is decreased. The flocculent structure of clay particles performs the same role as polymer network does. It reduced turbulence intensity and suppresses small eddies. Nevertheless, the concentration of clay suspension, at which the flocculent structure can form, is more than 100 000 times higher than polymer concentration, hence the viscous resistance is greatly enhanced due to high viscosity and yield stress of the suspension. For flow on smooth boundary, both viscous resistance and turbulent shear stress are important for energy consumption. The effect of increase in viscous resistance is often greater than that of the reduction in turbulent stress. Therefore, there is rare real drag reduction in this case.

On the other hand real drag reduction may occur in hyperconcentrated flow over rough bed because the energy consumption in this case is mainly due to turbulent stress and the increase in viscous resistance has minor influence. Figure 6.21 shows the average velocity over shear velocity of flows of clay suspension in an open channel with rough bed, U/U^*, as a function of the relative flow depth, H/D_{90}, where D_{90} was the representative size of rough elements on the bed (Rickenmann, 1990). All data were collected in the flows with the same slope. The points of clear water flow and the Nikuradse equation (with $k_s = D_{90}$) were plotted in the diagram for comparison. The average velocity with concentration $S_v = 4.7$-19.6% were higher than that of clear water flow. It means real drag reduction occurred in this case. But the points with concentration $S_v = 22.1\%$ fell

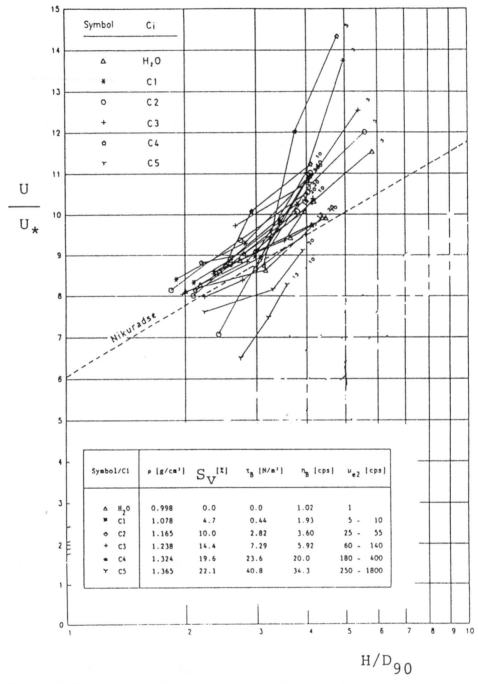

Figure 6.21. Flow resistance of clay suspension over a fixed rough bed.

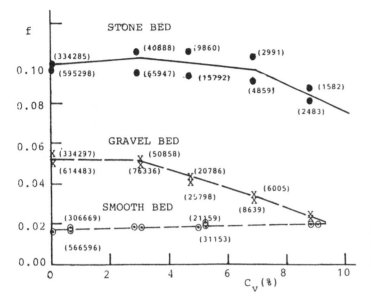

Figure 6.22. Friction factor as a function of clay concentration.

bellow the clear water line. Because this experiment was not conducted for studying drag reduction, no general conclusion and discussion on the phenomenon were presented.

Z. Wang (1993a) studied drag reduction in hyperconcentrated flow in an 24 m-long, 60 cm-wide flume with smooth, transitional and rough boundaries. The hydraulic smooth, transitional and rough boundary conditions for hyperconcentrated flow were achieved by covering the channel bed with steel plate, 12 mm gravels and 31.5 mm stones. The roughness for the gravel bed was 10 mm because the gravels were compactly arranged and glued on the bed and the protruding height was smaller than the gravels' diameter. The roughness of the stone bed was 31.5 mm because the stones were simply placed on the bed with certain mutual distances. It was observed that with gravel bed or stone bed the average velocity of flows of clay suspensions at high concentration was apparently higher and the flow depth was smaller than that of clear water flow at the same flow rate. Figure 6.22 presents the friction factor f as a function of the concentration of clay. The numbers in the parentheses are Reynolds number defined by Equation (6.14). The friction factor increased a little with increasing concentration in the flows on smooth bed. Nevertheless, it reduced following increase in concentration in the flows on the gravel bed as long as the concentration was higher than 3%. For the flows on the stone bed the drag reduction occurred when the concentration was higher than 4-7%. The mechanism of the drag reduction was also discussed in the literature (Z. Wang, 1993b).

6.3 INSTABILITY OF HYPERCONCENTRATED FLOW

6.3.1 *Phenomena of instability*

In the middle reach of the Yellow River and its tributaries in North-West China, a phenomenon so called 'river clogging' associated with hyperconcentrated flow sometimes occurs (see Chapter 9). Development of a hyperconcentrated flow into a series of roll waves was recorded at the Upper Lanxipo Station on the Black River – one of the tributaries of the Yellow River (Qian, 1980). As the sediment concentration rose up to 980 kg/m^3, the flow became unstable, and a series of waves formed along the river channel. The amplitude of the waves was 15-40 cm and the period was 8-10 minutes. The Froude number was less than 0.3. A Newtonian fluid cannot develop such waves at so low Froude number.

Another hyperconcentrated flow, with a concentration of sediment as high as 1220 kg/m^3, occurred in the Xiaoli River, North West China, in June, 1963. After the flood peak passed, the velocity of the hyperconcentrated flow reduced gradually and progressively until the flow stopped. The whole river was quiet as though it had frozen. This phenomenon is called 'river clogging'. Half an hour later the incoming discharge raised the surface slope and caused the hyperconcentrated fluid to flow again. This process repeated itself several times and the intermittent flow lasted 18.5 hours (Qian & Wan, 1983).

The similar phenomenon was also observed in the Yellow River. A flood occurred and the sediment concentration in the middle reach of the Yellow River reached 919 kg/m^3 in August, 1977. The flow of sediment-suspension transformed itself from a Newtonian turbulent flow into a non-Newtonian laminar flow. Waves 2-3 meters high appeared in the river. Gauging stations along the river recorded large scale fluctuations in both stage and discharge. The stage at the Zhaogou Station dropped 0.4 m in 2 hours (20:00-22:00) and then increased 1.8 m in 4 hours. At the Jiabu Station further downstream the stage dropped 0.95 m in 6 hours and then rose 2.84 m in 1.5 hours. Such dramatic changes in a river several kilometers wide were quite exceptional.

As debris flow develops from non-viscous debris flow into viscous one, it usually changes from a continuously turbulent flow into an intermittent laminar one. As shown in Figure 11.3, a viscous debris flow in the Jiangjia Gully developed into a series of waves and each wave lasted only 20-40 seconds. Between the waves the hyperconcentrated sediment mixture stopped flowing. Details of the phenomenon will be discussed in Chapter 11.

Life span of reservoirs depends mainly on the rate of sediment deposition. In order to sluice out sediment deposited in the reservoirs and extend their life spans, managers of reservoirs often empty the reservoir and allow incoming water to erode the deposited sediment. The flow often develops into fluctuating or intermittent flow (Guo et al., 1985). Figure 6.23 shows discharge and sediment concentration of mud flow in a model experiment that was designed to simulate

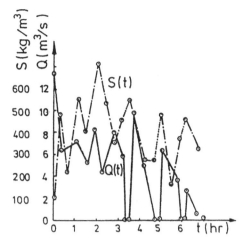

Figure 6.23. Fluctuation in discharge of mud flow.

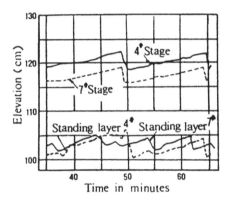

Figure 6.24. Fluctuation in stage in open channel mud flow.

the process of scouring cohesive sediment from a emptied reservoir (Wang, 1992). The time in the figure was in prototype scale. The flow was fluctuating and sometimes intermittent because the mud flow was non-Newtonian.

The instability of hyperconcentrated flow was experimentally studied by a few of researchers. Qian et al. (1980) made a flume experiment with clay mud and found that when the incoming discharge was small but stable, the stages at places with distances from the entrance of the flume rose gradually, and then suddenly fell down owing to a sudden acceleration of the fluid, as shown in Figure 6.24. The periodic variation repeated many times.

Engelund & Wan (1984) studied the mechanism of the instability of hyperconcentrated flow. The equipment they used for the experiment consisted of a 71 cm-long closed conduit with cross section 3.48 cm-wide × 1 cm-high and a surge tank with cross section 10 cm × 3.48 cm. Clay mud was fed into the surge tank at constant rate Q_{in} and then flowed out through the closed conduit. The stage

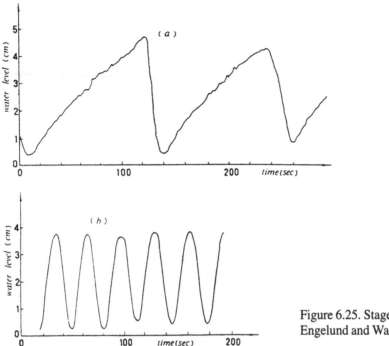

Figure 6.25. Stage fluctuation in Engelund and Wan's experiment.

in the surge tank (water head) was recorded automatically by a fluviograph. When the flow rate Q_{in} was high, the flow was stable and the stage in the surge tank was also stable. When Q_{in} was small, however, the flow was unstable. The measured water level in the tank fluctuated as shown in Figure 6.25. With even lower Q_{in} the stage fluctuated with longer period and greater amplitude, as shown in Figure 6.25(a). With higher Q_{in} the stage fluctuated with higher frequency and smaller amplitude. Engelund and Wan attributed the instability to thixotropy of the suspensions. They found that the montmorillonite suspensions exhibited thixotropy and the flow of it developed into fluctuating flow and the kaolinite suspensions did not exhibit thixotropy and the flow of it did not developed into fluctuations, thus they concluded that only the flow of the liquid which exhibits thixotropy can develop from stable flow into fluctuating or intermittent flow.

Wang et al. (1990) found that a hyperconcentrated flow could develop into fluctuating or even intermittent flow too, when the mixture did not exhibit thixotropy. They conducted experimental and theoretical studies on the instability of hyperconcentrated flow. The main mineral compositions of sediment they used were montmorillonite, calcite, kaolinite, quartz and mica. The sediment had a wide range of particle size with $d_{50} = 0.004$ mm and $(D_{84}/D_{16})^{0.5} = 4.7$. As concentration was higher than 50 kg/m^3 (or $S_v > 1.8\%$), the sediment suspension was a Bingham fluid. The yield stress and the rigidity coefficient increased with increasing concentration rapidly and no thixotropy was observed in the rheological measurements. The experiments were conducted in a recirculating flume 8.7 m long and 10 cm wide. The bed slope J could be adjusted from 0 to 0.2. Incoming discharge was recorded by means of a magnetic flow meter and the discharge at the outlet end of the flume was measured by using a gauging tank and a fluviograph. All flows in the experiments were laminar.

The mud flow remained stable if the incoming flow rate and mud depth were large enough, whereas the mud flow became unstable and a series of roll waves developed if the incoming flow rate and depth were small. Generally speaking, a slight fluctuation in velocity occurred first, and then some ripples appeared on the surface. Ripples grew into waves as they propagated downstreamwards, and more ripples formed at the same time. Sometimes waves grew so large that their maximum discharges were more than double the incoming discharge, and the residual mud stopped moving after waves passed.

A fully developed wave had the form shown in Figure 6.26. Figure 6.26a shows the streamlines of the mud flow as seen by viewer moving together with the wave. The wave always propagated more rapidly than the mud flow between waves did. The mud moved upward like a fountain under the extrusion of the wave as it was caught by a wave. Then it divided into two parts, one part flowed forward at a speed about $2U_w$ and formed a rolling front and the other part flowed at a speed less than U_w and gradually lagged behind the wave, in which U_w was the propagating speed of the wave. The developing stage of the wave occurred without the rolling front and all of the mud particles moved at a speed less than U_w. In the fully developed stage, the rolling front of the wave could break and fork as shown in Figure 6.26b.

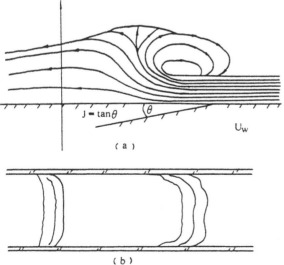

Figure 6.26. Profiles of mud wave.

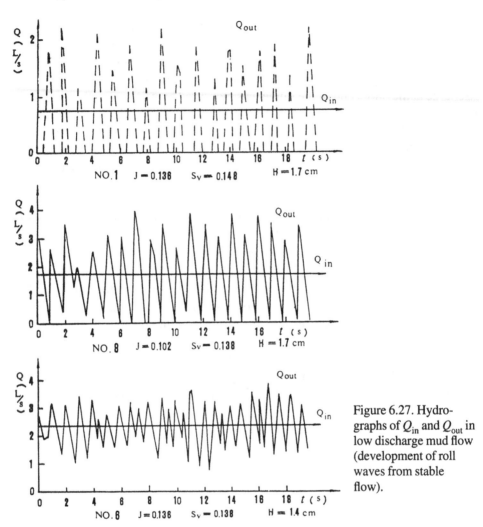

Figure 6.27. Hydrographs of Q_{in} and Q_{out} in low discharge mud flow (development of roll waves from stable flow).

Figure 6.27 shows the hydrographs of Q_{in} and Q_{out}. It shows that the mud flow at the outlet of the flume fluctuated and even developed into an intermittent flow (No.1), although the incoming discharge remained constant. The smaller the incoming discharge, the longer the time interval between the waves appeared to be. If Q_{in} was larger (No. 6), the flow was continuous but fluctuated.

The intermittent flow in Figure 6.27 was similar to viscous debris flow. In both cases, the slope was large and the depth of the flow was small. Viscous debris flow often develops into roll waves, e.g. debris flow waves occurred in the Jiangjia Gully, Yunnan Province, South China. The bed slope of the Jiangjia Gully was about 0.1, and the average mud depth was usually 10-60 cm as viscous debris flow takes place. The matrix of debris flow had a high yield stress. It changed from a continuous turbulent flow into an intermittent laminar flow and many surging waves developed as the concentration of solid material rose from several hundreds kg/m^3 to more than $1000\ kg/m^3$. Each wave lasted 20-40 seconds, and the time interval between the waves was about 20-100 seconds (Kang, 1985).

Neither roll waves nor fluctuations in velocity occurred when the incoming discharge was high or the bed slope was small in the experiments. Both conditions involved large flow depth. However, the flow in such a case could also become unstable if the upstream flow was disturbed. For example, if a mud wave was combined with the constant mud discharge by injecting mud from another feed tank, the wave could grow and induce additional waves during its propagation down the channel. This phenomenon is also attributed to the instability of hyperconcentrated flow.

At the beginning the flow in the whole flume was stable, the discharge measured at the outlet was the same as the discharge at the entrance. Then a mud wave was fed into the flow at the entrance from an additional feed tank. The wave grew as it propagated downstream under certain conditions. If V_o is the original volume of a wave, and V is the volume of the wave as it reached the outlet of the flume, the magnifying ratio, $\phi = V/V_o$, could be more than 6. The magnified wave was usually followed by a trough in which the discharge was smaller than the constant incoming discharge. Sometimes the trough was very deep and the discharge reduced to zero. The mud stopped flowing. Several seconds later the flow recurred and formed another wave. This phenomenon is similar to what is called 'river clogging'. Four waves induced by one disturbance wave were recorded in the experiment. Generally, the second wave was much lower than the first one and the second trough was much shallower than the first one, and the thirds further deteriorated. The increment of volume of the first wave was roughly equal to the defect volume of the first trough. The period of the wave was generally larger than 10 seconds, much longer than that of the roll waves in Figure 6.27.

The magnifying ratio, ϕ, depended little on the shape of the disturbance wave, but it depended strongly on the volume V_o. If V_o was too small, the wave gradually died out, instead of growing. Only if V_o was large enough could the magnifying ratio be larger than one. In the experiment, V_o varied from 1 liter to 10 liters, and the duration of the disturbance wave from 1.5 to 7 seconds. If the disturbance wave was tall and thin (duration short), it would be reduced and lengthened, and its volume could increase at the same time. If the wave was low and fat (duration long) but the volume V_o was not small, it would grow into a taller and larger wave.

Figure 6.28 shows the hydrographs of the incoming discharge (constant), the discharge of the disturbance wave Q_o, and the discharge at the outlet Q_{out}. The flow depths measured at $L = 1$ m (1 meter downstream from the entrance) and $L = 5.5$ m are also shown in the figure. Runs No. 26 and No. 31 had low incoming discharges (less than 1 l/s). Smaller disturbance waves grew into larger waves as

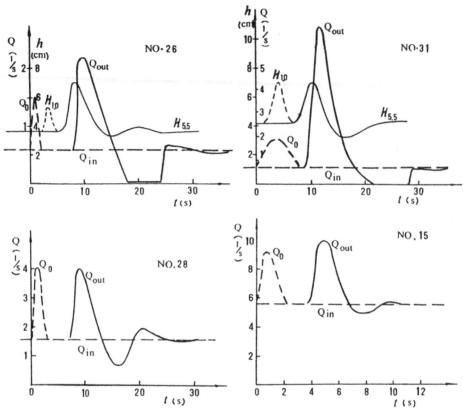

Figure 6.28. Development of disturbance waves.

they traveled toward the outlet of the flume, and flows stopped for 6.7 seconds after waves passed.

6.3.2 *Mechanism of instability*

Z. Wang (1993a) analyzed the mechanism of the instability of hyperconcentrated flow. For unsteady open channel flow, the one-dimensional continuity equation and equation of motion are

$$\frac{\partial H}{\partial t} + U\frac{\partial H}{\partial x} + H\frac{\partial U}{\partial x} = 0 \tag{6.37}$$

$$\frac{\partial U}{\partial t} + U\frac{\partial U}{\partial x} + g\frac{\partial H}{\partial x} = gJ - \frac{\tau_o}{\rho_m H} \tag{6.38}$$

where U is the average velocity, τ_o is the shear stress of the flow acting on the bed, or the resistance of the bed to the flow. With the method of the characteristics, the two partial differential equations are changed into two groups of normal differential equations in which one group follows the C_1-family of characteristic curves

$$\frac{dx}{dt} = U + \sqrt{gh} \tag{6.39}$$

$$\frac{d}{dt}(U + 2\sqrt{gH}) = gJ - \frac{\tau_o}{\rho_m H} \tag{6.40}$$

and another group follows the C_2-family of characteristic curves

$$\frac{dx}{dt} = U - \sqrt{gH} \tag{6.41}$$

$$\frac{d}{dt}(U - 2\sqrt{gH}) = gJ - \frac{\tau_o}{\rho_m H} \tag{6.42}$$

If a flow is steady and uniform, the velocity U and the depth H are constants, and the frictional resistance must then be equal to the tractive force, i.e.

$$gJ = \frac{\tau_o}{\rho_m H} \tag{6.43}$$

If a disturbance induces increments ΔU, $\Delta(2\sqrt{gH})$ and $\Delta(\tau_o/\rho_m H)$, then Equation (6.40) becomes

$$\frac{d}{dt}[(U + 2\sqrt{gH}) + \Delta U + \Delta(2\sqrt{gH})] = gJ - \frac{\tau_o}{\rho_m H} - \Delta\left(\frac{\tau_o}{\rho_m H}\right) \tag{6.44}$$

If one subtracts Equation (6.40) from Equation (6.44), the result is

$$\frac{d}{dt}[\Delta U + \Delta(2\sqrt{gH})] = -\Delta\left(\frac{\tau_o}{\rho_m H}\right) \tag{6.45}$$

Equation (6.45) is called disturbance equation along the C_1-family of characteristic curves. Similarly, the disturbance equation along the C_2-family of characteristic curves is

$$\frac{d}{dt}[\Delta U - \Delta(2\sqrt{gH})] = -\Delta\left(\frac{\tau_o}{\rho_m H}\right)_z \tag{6.46}$$

In laminar flow of a Bingham fluid, τ_o is given by Equation (6.1) in which du/dy is the velocity gradient near the bed. Figure 6.7 shows the general velocity distribution of a Bingham fluid in an open channel. The upper part is a flow plug in which the fluid flows at an uniform velocity u_p. Only in the zone near the bed does the velocity vary, from zero to u_p. The thickness of the layer is assumed d. Observations and measurements in the experiments suggested that d vary little as a wave passes, and the velocity gradient in the lower zone varied with fluctuation

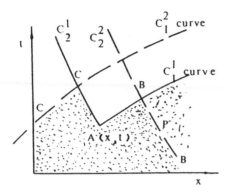

Figure 6.29. Characteristic curves on *x-t* plane.

of average velocity and depth. The velocity gradient is, therefore, roughly U/d, and

$$\tau_o = \tau_B + \eta \frac{du}{dy} \approx \tau_B + \eta \frac{U}{d} \tag{6.47}$$

Substituting the equation into Equation (6.45) one obtains

$$\frac{d}{dt}[(\Delta U + \Delta(2\sqrt{gH})] = -\Delta\left(\frac{\tau_B}{\rho_m H} + \eta \frac{U}{\rho_m H d}\right) \tag{6.48}$$

As shown in Figure 6.29, a disturbance that occurs at point A(x, t) in the *x-t* plane will propagate along the characteristic curves. For any point B on the C_1^1 characteristic curve that passes through the point A(x, t), a characteristic curve C_2^2 intersects the C_1^1 curve at point B. The dotted area is undisturbed. The initial disturbance has no effect on the area, and the velocity U and depth H in it remain constant. Integration of Equation (6.46) along the C_2^2 characteristic curve yields

$$\Delta U - \Delta(2\sqrt{gH}) = \int_{B'}^{B} -\Delta\left(\frac{\tau_o}{\rho_m H}\right) dt \tag{6.49}$$

Point P is always in the undisturbed area as it moves from B' to a point very near to B, and $\Delta(\tau_o/\rho_m H)$ is zero in the process of integration except at the point B. Since $\Delta(\tau_o/\rho_m H)$ is not infinite at B, Equation (6.49) yields

$$\Delta U - \Delta(2\sqrt{gH}) = 0 \tag{6.50}$$

or

$$\Delta U = \Delta(2\sqrt{gH}) \tag{6.51}$$

For a low discharge of mud flow, the average velocity is small. If the yield stress of the mud is large, the second term on the right hand side of Equation (6.48) is negligible, and Equation (6.48) can be rewritten as

$$\frac{d}{dt}\left[\sqrt{\frac{g}{H}}\,\Delta H\right] = -\frac{1}{2}\Delta\left(\frac{\tau_B}{\rho_m H}\right) = \frac{\tau_B}{2\rho_m H^2}\Delta H \tag{6.52}$$

In the process, Equation (6.51) and the following formulas

$$\Delta(2\sqrt{gH}) = \Delta H\frac{d}{dH}(2\sqrt{gH}) = \sqrt{\frac{g}{H}}\,\Delta H \tag{6.53}$$

$$\Delta\left(\frac{\tau_B}{\rho_m H}\right) = \Delta H\frac{d}{dH}\left(\frac{\tau_B}{\rho_m H}\right) = -\frac{\tau_B}{\rho_m H^2}\,\Delta H \tag{6.54}$$

have been used.

The integration of Equation (6.52) yields

$$\frac{\Delta H}{\Delta H_o} = e^{\frac{\tau_B}{2\sqrt{gH}\,H\rho_m}} \tag{6.55}$$

where ΔH_o is the initial disturbance in depth.

Equation (6.55) indicates that the initial disturbance ΔH_o will grow, and the larger the yield stress τ_o and the smaller the mud depth H, the faster the wave will grow. This trend is the reason that roll waves occurred in the experiments of small discharges and small depths. After a disturbance develops into a roll wave, the continuities of velocity and depth no longer hold, and the wave stops growing.

When the incoming discharge is large, the average velocity is high. If the rigidity coefficient η of the fluid is large, the first term on the right hand side of Equation (6.48) is negligible compared with the second term, and Equation (6.48) becomes

$$\frac{d}{dt}(\Delta U) = -\frac{1}{2}\Delta\left(\frac{\eta U}{\rho_m Hd}\right) \tag{6.56}$$

in which Equation (6.51) has been employed. Since

$$\Delta\left(\frac{\eta U}{\rho_m Hd}\right) = \frac{\partial}{\partial U}\left(\frac{\eta U}{\rho_m Hd}\right)\Delta U + \frac{\partial}{\partial H}\left(\frac{\eta U}{\rho_m Hd}\right)\Delta H$$

$$= \frac{\eta U}{\rho_m Hd}\left(\frac{\Delta U}{U} - \frac{\Delta H}{H}\right) = \frac{\eta}{\rho_m Hd}(1 - \text{Fr})\Delta U$$

Equation (6.56) can be rewritten as

$$\frac{d}{dt}(\Delta U) = -\frac{\eta}{2\rho_m Hd}(1 - \text{Fr})\Delta U \tag{6.57}$$

or after integration

$$\frac{\Delta U}{\Delta U_o} = e^{-\frac{1-Fr}{2\rho_m Hd}t} \tag{6.58}$$

where ΔU_o is the initial disturbance in velocity, $Fr = U/\sqrt{gH}$ is the Froude number.

Equation (6.58) proves that at high flow rate, as long as $Fr < 1$, the disturbance in velocity ΔU_o always decreases, hence, a large discharge of mud flow is stable.

Equations (6.55) and (6.58) give the results for the two extreme cases. In the general case, both terms on the right hand side of Equation (6.48) should be taken into account. Then we have

$$\frac{d}{dt}(\Delta U) = \frac{1}{2\rho_m H}\left[\frac{\tau_B}{\sqrt{gH}} - \frac{\eta}{d}(1 - Fr)\right]\Delta U \qquad (6.59)$$

If the fluid has a large yield stress and a small rigidity coefficient, the flow is unstable over a large range of discharges. The flow is liable to develop into roll waves. Whereas if τ_B is small and η is large, the flow is stable in most cases; only if the discharge and depth are very small is it possible to develop fluctuation in depth and discharge.

The foregoing discussion is based on the disturbance equation along the C_1-family of characteristic curves. The initial disturbance at point $A(x, t)$ propagates along the C_2-family of characteristic curves (see AC curve in Figure 6.29) too. Integration of Equation (6.45) along the characteristic curve C_1^2 yields

$$\Delta U = - \Delta(2\sqrt{gH}) = -\sqrt{g/H}\Delta H \qquad (6.60)$$

Substituting the equation into

$$\frac{d}{dt}[\Delta U - \Delta(2\sqrt{gH}] = -\Delta\left[\frac{\tau_B}{\rho_m H} + \frac{\eta U}{\rho_m H d}\right] \qquad (6.61)$$

one obtains the result

$$\frac{d}{dt}(\Delta U) = - \frac{1}{2\rho_m H}\left[\frac{\tau_B}{gH} + \frac{\eta}{d}(1 - Fr)\right]\Delta U \qquad (6.62)$$

The equation shows that no matter how large the discharge and the depth are, the τ_B-term or the η-term dominates the resistance; hence, the disturbance wave always declines along the C_2-family of characteristic curves.

The analysis proved that the instability of hyperconcentrated flow is attributed to the yield stress of the mixture. The viscosity or rigidity coefficient of the fluid is a factor to resist the instability. Fluctuation in discharge and depth may occur as long as the hyperconcentrated flow transforms from Newtonian into non-Newtonian fluid and exhibits yield stress. The greater the yield stress and the smaller the rigidity coefficient, the more unstable the hyperconcentrated flow. At low flow rate the yield stress dominates the resistance and the flow is liable to develop into unstable flow or even intermittent flow; whereas at high flow rate the rigidity coefficient dominates the resistance the flow remains stable.

REFERENCES

Engelund, F. & Z. Wan 1984. Instability of hyperconcentrated flow. *J of Hydraulic Engineering*, Vol. 110(3): 219-233.

Fang, Z. & G. Hu 1984. Some characteristics of flow with hyperconcentration in the Bajiazui Reservoir (in Chinese). *J. of Sediment Research*, No. 1: 73-75.

Guo Z., B. Zhou, L. Ling & D. Li 1985. The hyperconcentrated flow and its related problems in operation at Hengshan Reservoir. *Proc. Intern. Workshop on Flow at Hyperconcentrations of Sediment, IRTCES, Sep. Beijing, China.*

Hedstrom, B.O.A. 1952. Flow of plastic materials in pipes. *Indus. Engin. Chem.*, Vol. 44: 651-656.

Hishinko, P.H. 1951. *Mud Hydraulics* (in Chinese). Translated by Yuan & Chen (1957). Oil Industry Publishing Inc.

Hou, H. & X. Yang 1983. Effect of fine sediment on the drag reduction in muddy flow. *Proc. of the Second Intern. Sympos. on River Sedimentation*, pp. 47-80.

Hua, J. 1986. A study on resistance in sediment transportation in pipe lines and relative problems (in Chinese). Master thesis, Inst. of Water Conservancy and Hydroelectric Power Research.

Kang, Z. 1985. *Characteristics of the flow patterns of debris flow at Jiangjia Gully in Yunnan* (in Chinese). Memoirs of Lanzhou Institute of Glaciology and Crypediology, No.4.

Qian, N. 1980. *A preliminary study of the mechanism of hyperconcentrated flows in North West China* (in Chinese). Academic Reports on Sediment Problems in The Yellow River, No. 4.

Qian, N. & Z. Wan 1983. *Mechanics of Sediment Movement* (in Chinese). Chinese Science Press.

Qian, N. (N. Chien) & Z. Wan 1985. A critical review of the research on the hyperconcentrated flow (in Chinese). *J. of Hydraulic Engineering*, No. 5: 27-34.

Qian Y., W. Yang, W. Zhao, X. Chen, L. Zhang & W. Xu 1980. Basic characteristics of flow with hyperconcentration of sediment. *Proc. Intern. Symp. on River Sedimentation, Guahua Press, Mar. China.*

Ren, Z. 1985. Experimental study on resistance of flow with hyperconcentration of sediment in open channels and pipes. *Proc. of Intern. Workshop on Flow at Hyperconcentration of Sediment IRTCES, II-02.*

Rickenmann, D. 1990. Bedload transport capacity of slurry flows at steep slopes. *Versuchsanstalt für Wasserbau, Technischen Hochschule Zürich, Mitteilungen* 103.

Song, T., Z. Wan & N. Qian (Ning Chien) 1986. The effect of fine particles on the two-phase flow with hyperconcentration of coarse grains (in Chinese). *J. of Hydraulic Engineering*, No. 4: 1-10.

Toms, B.A. 1948. Some observations on the flow of linear polymer solutions through straight tubes at large Reynolds numbers. *Proc. of First Intern. Congr. on Rheology*, Vol. II: 135-141.

Wan, Z., Y. Qian, W. Yang & W. Zhao 1979. Laboratory study on hyperconcentrated flow (in Chinese). *People's Yellow River*, pp.53-65.

Wang, H. & X. Fei 1989. The fluctuation of Bingham shear stress of hyperconcentrated flow in flowing conditions. *Proc. of 4th Intern. Symp. on River Sedimentation*, pp. 198-205.

Wang, M., Y. Zhan, J. Liu, W. Duan & W. Wu 1983. An experimental study on turbulence characteristics of flow with hyperconcentration of sediment. *Proc. of 2nd Intern. Symp. on River Sedimentation, Chinese Water Resources and Electric Power Press*, pp. 36-45.

Wang, Z. 1992. Model test of non-Newtonian fluid flow. *Proc. 5th Intern. Symp. on River Sedimentation, IWK and IRTCES, Karlsruhe, Germany.*

Wang, Z. 1993a. A study on debris flow surges, *Hydraulic Engineering'93*, Volume 2, American Society of Civil Engineers, New York, pp. 1616-1621.

Wang, Z. 1993b. *Structural features of turbulent flow of clay suspensions over rough boundary.* Research Report, Inst. f. Wasserbau und Kulturtechnik, University of Karlsruhe.

Wang, Z. & N. Qian (Ning Chien) 1984. Experimental study on the physical properties of sediment suspension with hyperconcentration (in Chinese). *J. of Hydraulic Engineering*, No. 4: 1-10.

Wang, Z., B. Lin & X. Zhang 1990. Instability of non-newtonian fluid flow (in Chinese). *Mechanica Sinica*, No. 3: 266-275.

Wang, Z., Y. Ren & X. Wang 1992. Turbulent structure of Bingham fluid open channel flow (in Chinese). *J. of Hydraulic Engineering*, No. 12: 9-17.

Xu, Y. & Z. Wan 1985. The transport and utilization of hyperconcentrated flow in Luohui Irrigation District. *Proc. of Intern. Workshop on Flow at Hyperconcentration of Sediment, IRTCES*, pp. III-6.

Yang, W. & W. Zhao 1983. An experimental study of the resistance to flow with hyperconcentration in rough flumes. *Proc. of 2nd Intern. Symp. on River Sedimentation, Water Resources and Electric Power Press*, pp.45-55.

CHAPTER 7

Hyperconcentrated flow with cohesionless particles

7.1 HYPERCONCENTRATED LAYER

Suspended load is supported by diffusive force in turbulent flow. The diffusive force depends on concentration gradient and turbulence intensity. Fine sediment requires smaller diffusive force and coarse sediment requires larger diffusive force to be suspended. Therefore, the coarser the sediment, the greater the concentration gradient in a flow with a certain turbulence intensity. There often exists a hyperconcentrated layer near the bed if an open channel flow carries coarse sediment. Sediment in the layer moves either as suspended load or as bed load. Because turbulent eddies are produced in a zone near the bed, the existence of the hyperconcentrated layer is bound to affect generation and transmission of eddies, and thus changes velocity distribution.

7.1.1 The phenomenon of hyperconcentrated layer

As early as in the fifties Einstein & Chien (1955) studied hyperconcentrated layer near bed and its effect on velocity distribution. The study was conducted in a recirculating tilting flume whose bed was roughened by gluing sand on it. Three kinds of uniform sediment with diameters 1.3 mm, 0.94 mm and 0.274 mm were used. Figure 7.1 shows the measured concentration distribution, where H and y are depth and elevation from the bed, and S the sediment concentration in kg/m^3. In a zone near the bed, $(H - y)/y > 10$, S reached 60-600 kg/m^3.

Figure 7.2 shows sediment distribution measured by Yano & Daido (1969). They used plastic particles with diameter 7.43 mm and specific weight 1.44 g/cm^3. About 20-30% of the depth is a hyperconcentrated layer with volume concentration $S_v > 10\%$.

Hyperconcentrated layer also occurs in hydrotransport of granular materials in pipelines. Figure 7.3 shows sediment concentration distribution by Newitt et al. (1962). There exists a hyperconcentrated layer in the lower half of the flow with

129

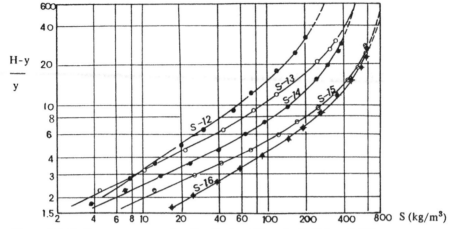

Figure 7.1. Sediment concentration distribution (Einstein & Chien, 1955).

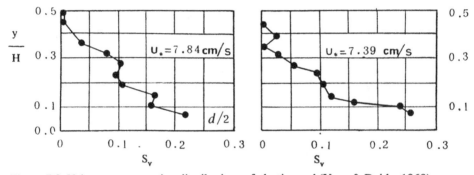

Figure 7.2. Volume concentration distributions of plastic sand (Yano & Daido, 1969).

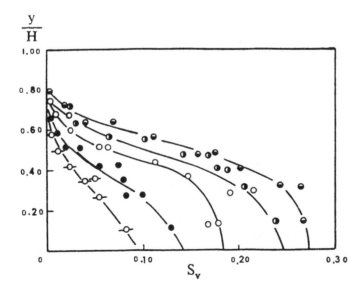

Figure 7.3. Concentration distribution of coarse sand in pipe flow (Newitt et al., 1962).

volume concentration S_v = 10-25%, while in the upper half of the flow the concentration is nearly zero.

7.1.2 *Energy structure of hyperconcentrated layer*

Chien & Wan (1965) analysed Einstein and Chien's data and presented an energy structure of flow with hyperconcentrated layer. Figure 7.4 shows velocity, shear stress and sediment concentration profiles in the flow with hyperconcentrated layer and corresponding ones in the flow of clear water. The velocity in the flow with hyperconcentrated layer is lower in a zone near the bed, $y/H < 0.19$, and higher in the zone $y/H > 0.19$ than those in clear water flow. The shear stress distributions in the two cases are about the same. By using the distributions of velocity and shear stress, distributions of energy indexes W_b, W_s and W_t are presented in Figure 7.5, where W_b is the energy supplied from potential energy in unit volume and unit time, W_s is the energy dissipated locally, and W_t is the energy transmitted from the upper part to the lower part. These energy indexes can be expressed as follows

$$W_b = -u\frac{d\tau}{dy} \tag{7.1}$$

$$W_s = \tau\frac{du}{dy} \tag{7.2}$$

$$W_t = W_b - W_s = -\frac{d(\tau u)}{dy} \tag{7.3}$$

In clear water flow energy dissipates mainly in an extremely thin layer near the bed, W_s reduces rapidly and vanishes following increase in distance from the bed. In the flow with hyperconcentrated layer, W_s is greater in most part of the flow

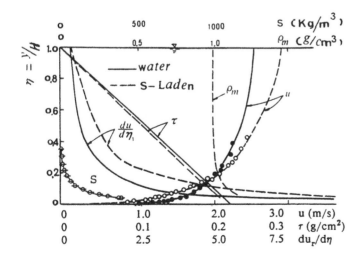

Figure 7.4. Distributions of main parameters of clear water and sediment-laden flow (Chien & Wan, 1965).

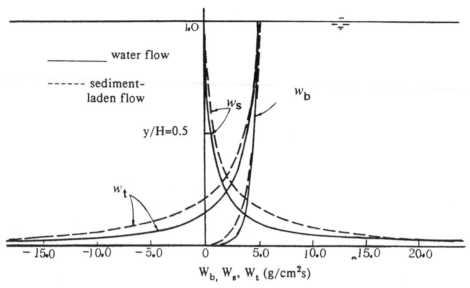

Figure 7.5. Energetic structures of clear water flow and sediment-laden flow (Chien & Wan, 1965).

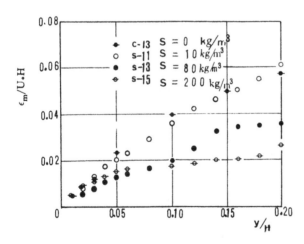

Figure 7.6. Comparison of the eddy viscosity coefficients between clear water flow and sediment-laden flow.

because energy is consumed not only by friction on the boundary but also by carrying sediment.

In Figure 7.6 the dynamic eddy viscosity ε_m in clear water flow and flow with hyperconcentrated layer are compared, in which

$$\varepsilon_m = \frac{\tau}{\rho \dfrac{du}{dy}}$$

(7.4)

or

$$\frac{\varepsilon_m}{U_* H} = \frac{1 - \eta}{\dfrac{du_r}{d\eta}} \tag{7.5}$$

where

$$u_r = u / U_*, \eta = y/H \tag{7.6}$$

The dynamic eddy viscosity in sediment-laden flow is apparently less than that in clear water flow and the higher the concentration, the smaller the eddy viscosity.

7.1.3 *Effect of hyperconcentrated layer on velocity distribution*

Vanoni (1946) concluded that the velocity distribution in sediment-laden flow still follows the logarithmic formula, but the von Karman constant is less than 0.4. Sutton (1953) studied velocity distribution of wind near the ground and found the velocity gradient du/dy follows the following formula:

$$\frac{du}{dy} = a y^{-\beta} \tag{7.7}$$

where a is a constant and β varies with density gradient of air,

$$\beta > 1, \text{ if } d\rho/dy > 0$$
$$\beta = 1, \text{ if } d\rho/dy = 0$$
$$\beta < 1, \text{ if } d\rho/dy < 0 \tag{7.8}$$

where ρ is the density of air.

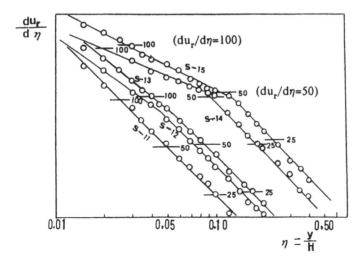

Figure 7.7. Distribution of velocity gradient.

Chien & Wan (1965) compared the velocity distribution in hyperconcentrated layer with above formula and expounded that the effect of negative concentration gradient in hyperconcentration layer is similar with the effect of negative density gradient of air on wind velocity distribution. With the definition of u_r and η, Equation (7.7) is rewritten as

$$\frac{\mathrm{d}u_r}{\mathrm{d}\eta} = a'\eta^{-\beta} \tag{7.9}$$

where $a' = AH^{1-\beta}/U_*$. Figure 7.7 shows distribution of $\mathrm{d}u_r/\mathrm{d}\eta$ measured by Einstein and Chien. In the most part of the flow where the concentration and concentration gradient are small, the velocity gradient follows straight lines at angle $-45°$, or $\beta = 1$ in Equation (7.9). In hyperconcentrated layer where the negative concentration gradient and thus the negative density gradient are great the measured points follows flatter lines, that suggests β-value less than 1.

7.2 TWO-PHASE TURBULENT FLOW WITH HYPERCONCENTRATION OF COHESIONLESS SEDIMENT

7.2.1 *Wash load and bed material load in hyperconcentrated flow*

Suspended sediment is classified as bed material load and wash load. The criterion for differentiating wash load from bed material load used to be taken as D_{b5} or D_{b10}, which represent diameters pertaining to 5% or 10% in the granulometric curve of sediment on the bed. In hyperconcentrated flow, however, suspended sediment of much larger diameter than D_{b10} can also be transported for a long distance without deposition and exchange with sediment on the bed. It is essentially washed through the channel and should belong to the category of wash load. It is mentioned in Chapter 2 that the Rouse number Z has been suggested to be employed as a new criterion for differentiating wash load from bed material load. The development of the new criterion started from a solution of the diffusion equation (Wang & Dittrich, 1992).

In the following, a two dimensional open channel flow with suspended sediment is considered. Suspended particles move together with water in the flow direction, fall towards the bed under the action of gravitation and diffuse in the opposite direction of the concentration gradient. The amount of suspended sediment transported through a unit area of a section in the flow field with a normal unit vector $\vec{n}$ is given by $\vec{q}\cdot\vec{n}$, in which

$$\vec{q} = (u\vec{i} - \omega\vec{j})S_v - \varepsilon_s \nabla S_v \tag{7.10}$$

where ∇S_v is the concentration gradient vector, $\vec{i}$ and $\vec{j}$ are the unit vectors in the flow direction (x-coordinate) and the vertical direction (y-coordinate), respectively, ω is fall velocity of particles. The mass conservation equation gives

$$\partial S_v / \partial t + \nabla \cdot \vec{q} = 0 \tag{7.11}$$

Assuming u independent of x, ω independent of y and $\partial S_v / \partial t = 0$, Equation (7.11) can be rewritten as

$$u\frac{\partial S_v}{\partial x} = \omega\frac{\partial S_v}{\partial y} + \frac{\partial \varepsilon_s}{\partial y}\frac{\partial S_v}{\partial y} + \varepsilon_s\frac{\partial^2 S_v}{\partial y^2} \tag{7.12}$$

Equation (7.12) is the general diffusion equation. The boundary conditions can be formulated as follows:

1. On the surface $\vec{q} \cdot \vec{j} = 0$, or

$$-\left(\omega S_v + \varepsilon_s\frac{\partial S_v}{\partial y}\right) = 0, \text{ at } y = H \tag{7.13}$$

2. The sediment concentration near the bed is determined by the bed load concentration and is constant in steady flow,

$$S_v(x, y_o) = S_{vo} \tag{7.14}$$

3. The sediment concentration at the entrance is assumed to be uniformly distributed

$$S_v(0, y) = S_{vi} \tag{7.15}$$

The sediment-diffusion coefficient ε_s is assumed to be proportional to the turbulent momentum-diffusion coefficient, or the eddy viscosity, ε_m (see Equation (7.5)). Using the Prandtl-von Kármán velocity defect law, Equation (7.4) can be written as

$$\varepsilon_m = \kappa U_* y(1 - y/H) \tag{7.16}$$

where U_* is the shear velocity, κ the von Kármán constant. Equation (7.16) is not valid close to the boundary $y = H$ because diffusion of momentum and sediment at the surface still exist. Therefore, it is modified as follows

$$\varepsilon_m = \begin{array}{l} \kappa U_* y(1 - y/H), y < aH \\ \kappa U_* Ha(1 - a), aH < y < H \end{array} \tag{7.17}$$

where a is a constant and found to be equal to 0.67 by Ismail (1952) and to 0.7 by Wang & Qian (1984). The numerical solutions of Equation (7.12) are computed with an explicit difference scheme.

Figure 7.8 gives the computed concentration profiles at different distances from the entrance under the boundary conditions $S_{vo} = 0.2$ and $S_{vi} = 0$, 0.2 and 0.4 with a Rouse number $Z = 2$. The results show that the concentration distributions are the same and stable for different incoming sediment concentrations S_{vi},

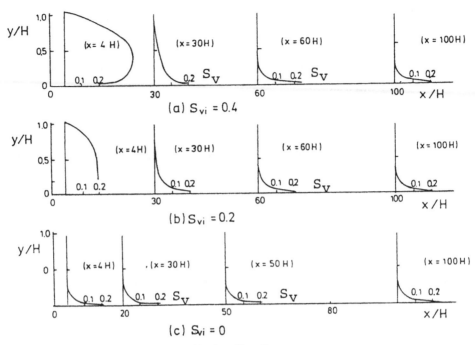

Figure 7.8. Calculated sediment distribution ($Z = 2$).

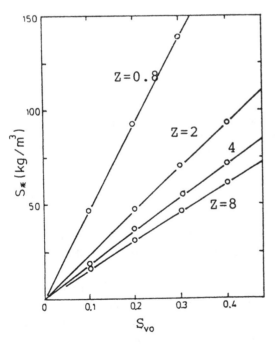

Figure 7.9. Sediment-carrying capacity S_* as a function of S_{vo} and Z.

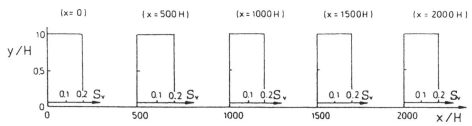

Figure 7.10. Calculated sediment distribution ($Z = 0.01$).

as long as $x > 60H$ which is the equilibrium distance for adjustment of sediment distribution. The results for $Z = 8$, 4, 2 and 0.8 are similar, although the equilibrium distances are different. In all these cases, the sediment concentration at distances larger than $400H$ is determined by the concentration of the bed load zone S_{vo} and the value of Z, and is independent of the incoming sediment concentration. Figure 7.9 shows S_*, the average sediment concentration in equilibrium, as a function of S_{vo} and Z from the numerical solution, S_* represents sediment transporting capacity of the flow. In general, S_* increases with increasing S_{vo} and decreasing Z. If $Z > 4$, however, the Rouse number has little influence on the sediment transporting capacity, because sediment moves mainly as bed load instead of suspended load at so great Z values. Therefore the total sediment transport capacity depends only on rate of bed load transport.

However, as Z is very small, the results are completely different. Figure 7.10 shows the concentration profiles for $Z = 0.01$. Very small Z values imply weak settling and strong turbulent diffusion. This is the case either the sediment is fine or the turbulence intensity is high. It is also such a case that the liquid phase consists of clay suspension. As shown in Figure 7.10, for small Rouse numbers the concentration profiles are uniform and only depend on the incoming concentration S_{vi}.

From the results presented above, it follows that a strong exchange between suspended sediment and bed material exists if Z is large. In contrary, the suspended sediment is essentially washed through the channel and no exchange with bed material will occur if Z is small. Thus, we can conclude that the sediment-laden flow with large Z values should be referred to bed-material load and that with small Z values to the category of wash load. According to the computed results, the transition from wash load to bed-material load is in the range of $Z = 0.06$ to $Z = 0.1$, i.e. the suspended sediment is wash load if $Z < 0.06$ and bed-material load if $Z > 0.1$. With the new criterion wash load and bed-material load are mutually transformable. During hyperconcentrated floods in the Yellow River and its tributaries in Northwest China, the high concentration of sediment and the existence of fine particles largely reduces the fall velocity of coarse particles. Consequently, the value of Z for coarse sediment can be smaller

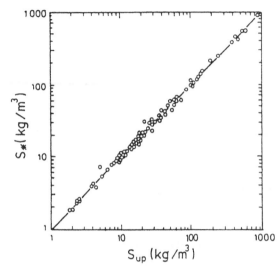

Figure 7.11. S_* versus S_{up} for wash load ($Z < 0.06$).

than 0.06. In this case, coarse particles with diameters larger than 0.1 to 0.2 mm can be transported as wash load.

Figure 7.11 shows the average weight concentrations S (kg/m³) measured in the Yellow River and its tributaries in the case $Z < 0.06$ versus the corresponding values measured at upstream gauging station S_{up}. The values of S_{up} were taken a time period T ahead, where T (= L/U) equals the distance between the two stations (more than 100 km) over the average velocity. The results prove that for sediment with $Z < 0.06$ the concentration does not change even after a travel over 100 km, and thus

$$S = S_{up} \tag{7.18}$$

In the middle reach of the Yangtze River the average depth is about 4 m to 30 m, the average river slope is about 0.015% to 0.1% and the average sediment concentration is lower than 1 kg/m³. The critical diameter for wash load, according to the new criterion, is 0.06 mm to 0.15 mm depending on the shear velocity U_*. These values coincide well with the empirical values from investigations for a great project on the river (the Three Gorges Project) (YVPO, 1985).

7.2.2 *Velocity distribution of two-phase flow with hyperconcentration of cohesionless sediment in pipe lines*

There are rare hyperconcentrated two-phase flows with pure cohesionless sediment in nature. Hydrotransport of solid materials in pipelines, nevertheless, often involves hyperconcentrated two-phase flow consisting of cohesionless particles. Many studies on distributions of velocity and turbulence intensity in hyperconcentrated two-phase flow in closed conduit have been reported. Silin et al. (1969)

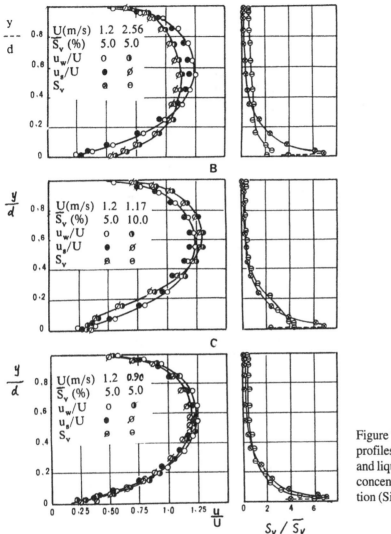

Figure 7.12. Velocity profiles of solid phase and liquid phase and concentraion distribution (Silin et al., 1969).

measured velocities of fluid and solid phases in flows of water with 0.46 mm suspended sediment as shown in Figure 7.12, in which d is the diameter of the pipe, u_w and u_s are velocities of the liquid and that of the solid, respectively, S_v is average concentration by volume. The distributions of u_w and u_s are very close to each other but all measured u_s is slightly smaller than the corresponding u_w. Figure 7.13 shows distributions of root mean square of fluctuating velocities of the two phases and the same conclusion holds, i.e. the measured points for the two phases are very close to each other and the values for the solid phase are a little bit smaller than that for the liquid phase.

Deriving from assumptions on eddy viscosity, Ayukawa (1970) presented a

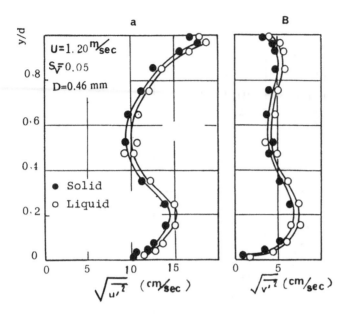

Figure 7.13. Distributions of the root mean square of fluctuating velocities of the solid and liquid phases (Silin et al., 1969).

velocity distribution of flow with heterogeneously suspended sediment as a function of concentration distribution. Shear stress in the two-phase flow is expressed as follows:

$$\tau = \rho\varepsilon_m \frac{du}{dy} + (\rho_s - \rho)\varepsilon_s \frac{d(S_v u)}{dy} \tag{7.19}$$

where ρ is the density of water, ρ_s the density of the solid. The first term in the equation represents shear stress owing to turbulence and the second one represents mean momentum flux due to turbulent diffusion of solid particles.

Assume the maximum velocity occurs at $y = y_m$, y is the distance from the bottom wall, distribution of the shear stress is given by:

$$\tau = \tau_{ol}\left(1 - \frac{y}{y_m}\right) \tag{7.20}$$

and

$$\tau = \tau_{ou}\left(1 - \frac{H - y}{H - y_m}\right) \tag{7.21}$$

where τ_{ol} and τ_{ou} denote the shear stresses on the bottom and the top walls, H is the height from the bottom wall to the top wall. If (1) distribution of the eddy viscosity ε_m in the two-phase flow and water flow is the same; and (2) the following relationships exist:

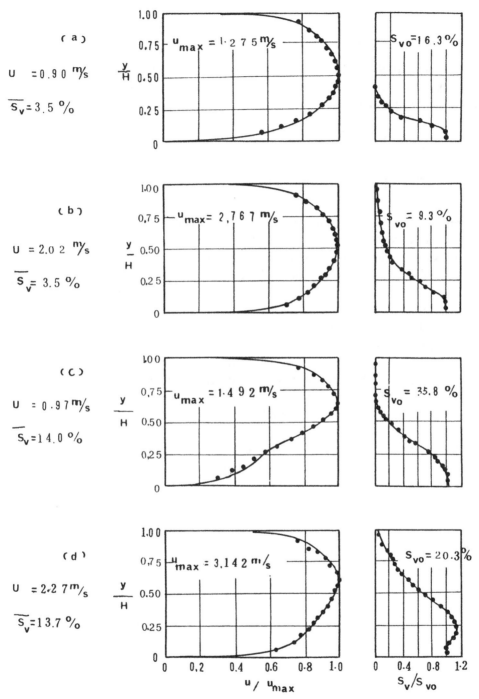

Figure 7.14. Velocity profiles and volume concentration profiles (Ayukawa, 1970).

$$\frac{\varepsilon_m}{U_{*L}y_m} = k_L\frac{2\varepsilon_{mW}}{U_{*w}H}; \frac{\varepsilon_m}{U_{*u}(H-y)} = k_u\frac{2\varepsilon_{mW}}{U_{*w}H} \qquad (7.22)$$

where subscript W indicates flow of clear water, L and u denote the lower half and the upper half of flow in closed conduit, the following velocity profile for two phase flow is presented:

$$\frac{u}{u_{max}} = \frac{1 + \beta\left(\frac{\rho_s}{\rho} - 1\right)S_{vm}}{1 + \beta\left(\frac{\rho_s}{\rho} - 1\right)S_v} \frac{u_W}{u_{Wmax}} \qquad (7.23)$$

where S_{vm} is volume concentration at $y = y_m$, u velocity of two-phase flow and u_W velocity of water flow on the same conditions, and $\beta = \varepsilon_s/\varepsilon_m$.

Ayukawa made an experiment in a horizontal, straight and transparent duct of square cross section of 40×40 mm^2. He used water and polycarbonate pellets, of diameter 3.06 mm and specific gravity 1.165, as the fluid phase and the solid phase. Figure 7.14 shows measured concentration and velocity profiles and calculated velocity profile with Equation (7.23). The formula agrees well with the measured velocity profiles.

Equation (7.23), however, does not give a proper velocity profile, but only gives a relationship between the velocity profiles of sediment-laden flow and clear water flow as a function of concentration distribution. Applicability of the equation is limited because concentration profile is not easier to be measured than velocity profile.

A distribution of eddy viscosity in pipe flow is given by Equation (7.16) with y_m replacing H. The equation represents two parabolas as shown in Figure 7.15a. It has been discussed in Chapter 2 that the distribution of eddy viscosity should be

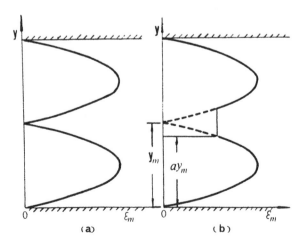

(a)   (b)

Figure 7.15. Distributions of eddy viscosity ε_m in pipe flow.

modified because turbulence intensity in the center of the pipe flow is not zero. It is modified, like Figure 7.15b, with mathematical form of Equation (7.17).

With the modified distribution of eddy viscosity, the following velocity distribution is derived:

$$\frac{u_{\max} - u}{U_*} = \begin{cases} \dfrac{1 - a}{2a\kappa} + \dfrac{1}{\kappa}\ln\dfrac{ay_m}{y}, & y \leq ay_m \\[2ex] \dfrac{1}{2a(1 - a)\kappa}\left(1 - \dfrac{y}{y_m}\right)^2, & ay_m < y < y_m \end{cases} \tag{7.24}$$

Wang & Qian (1984) conducted a series of experiments in a horizontal closed conduit with rectangular cross section 180×100 mm^2. Water and cohesionless sediment of diameter 0.15 mm and specific gravity 2.64 were used as the liquid phase and the solid phase. The average velocity and the average concentration were in the ranges 0.5-5 m/s and 0-1000 kg/m^3. The measured velocity distribution of flow with clear water followed Equation (7.24) very well and the constant κ was about 0.4. In the flows with hyperconcentration of sediment, however, the measured velocity profiles deviates from Equation (7.24) in the lower zone because suspended sediment affects eddy diffusivity through the following ways.

(a) *Density gradient.* The non-uniform distribution of sediment results in a negative density gradient. As a lump of the two-phase mixture comes up from the lower part where the density of the mixture is higher, turbulent energy is dissipated to overcome the gravitation because the buoyancy force in the upper flow is less than the weight of the upcoming lump. On the other hand, as a lump from upper part moves downward, turbulent energy is also dissipated to overcome the buoyancy force since the buoyancy force in the lower part is greater than the weight of the downcoming lump. Turbulence intensity of the flow is thereby smaller than that of clear water flow.

(b) *Lifting force.* Solid particles in a flow field with velocity gradient is subjected to action of a lifting force. The larger the solid particle and the larger the velocity gradient, the greater the lifting force. The effect of the lifting force is significant near the bed where the velocity gradient and particles concentration is high and negligible in most part of the flow. Such an effect strengthens the vertical momentum exchange and reduces velocity gradient so that it augments ε_m.

The effect of density gradient changes the constant κ and the effect of the lifting force changes the velocity distribution near the bed. Assuming the increment of ε_m is proportional to $D^2(du/dy)$, Wang and Qian obtained the following equation:

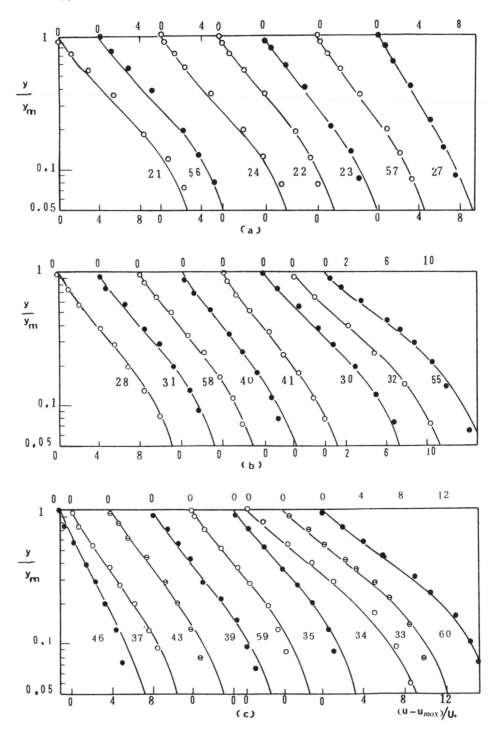

Figure 7.16. Comparison of Equation (7.25) with measured velocity profiles of hyperconcentrated flows in a closed conduit (Wang & Qian, 1984).

$$\frac{u_{max} - u}{U_*} = \frac{\dfrac{1-a}{2\kappa a} + \dfrac{1}{\kappa} ln \dfrac{ay_m}{y} - \dfrac{5}{\kappa^3} \dfrac{D^2}{y^2} \left[\dfrac{y_m + 2y}{y_m} - \dfrac{1 + 2a}{a^2} \dfrac{y^2}{y_m^2} \right], y \leq ay_m}{\dfrac{1}{2\kappa a (1-a)} \left(1 - \dfrac{y}{y_m} \right)^2, ay_m < y \leq y_m}$$

(7.25)

where $a = 0.7$ and κ is no longer 0.4. Figure 7.16 gives a comparison of the formula with the measured velocity profiles. The measured points follow the curves rather closely.

A Richardson number in the following form was adopted

$$\text{Ri}_m = \frac{gy_m}{U_*^2} \frac{\rho_{ma} - \rho_{mb}}{\rho_m}$$

(7.26)

where ρ_{ma} and ρ_{mb} are the densities of the mixture at the elevations $(y_m - y)/y = 20$ and $(y_m - y)/y = 0.04$, respectively, ρ_m is the average density of the mixture. κ is plotted in Figure 7.17 against Ri_m and the average volume concentration S_v. Measured points in hyperconcentrated flows in the Wuding River at Dingjiagou Station and laboratory study by Einstein & Chien (1955) are also plotted in the diagram. Except a few scattering points, the Karman constant in hyperconcen-

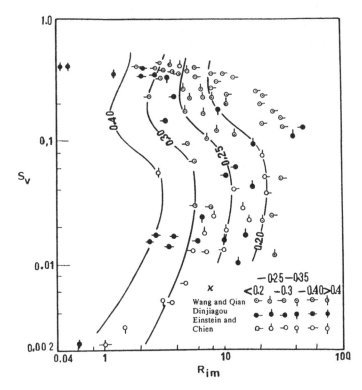

Figure 7.17. The variation of κ with S_v and Ri_m.

trated flows in pipeline, river and flume obeys the same rule. The Karman constant decreases monotonously with increasing Ri_m, which indicates that the density gradient reduces momentum diffusion and increases velocity gradient. κ also varies with concentration S_v but in a more complicated way. It seems that at $S_v = 0.1\text{-}0.2$ (or $S = 260\text{-}530 \text{ kg/m}^3$), the Karman constant reduces to the minimum.

7.2.3 *Concentration distribution of sediment in hyperconcentrated pipe flow*

It is well established that concentration profile of suspended sediment in open channel flow follows the Rouse function:

$$\frac{S_v}{S_{vb}} = \left[\frac{H-y}{y} \frac{b}{H-b} \right]^z \tag{7.27}$$

where H is the depth of the flow, S_{vb} is the concentration at the point $y = b$.

Sediment concentration distribution in horizontal closed conduit with low average concentration ($S < 31 \text{ kg/m}^3$) still follows Rouse function, as proved by Ismail (1952). Nevertheless, concentration distribution in hyperconcentrated flow in closed conduit is quite different from Rouse function. All available experimental data, such as Newitt et al.'s (Figure 7.3), Ayukawa's (Figure 7.14), Stevens' (1972, Figure 7.18), and Charles' (1970, Figure 7.19) indicate that the concentration distribution is distinctly different from those in open channel flow. With sufficiently high concentration, a turning point appears in the central zone of the pipe flow and the concentration distribution curve transforms from concave downward to convex upward. With further higher concentration, positive concentration gradient may be observed in the lower part of the flow. Because of the great complexity, it is difficult to formulate concentration distribution in hyperconcentrated pipe flow.

Wang & Qian (1984) analysed the diffusion process and found that the diffusion equation consists of two parts in closed conduit flow. For demonstration, let us consider a sediment-laden flow in a horizontal pipe line with rectangular

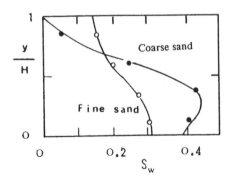

Figure 7.18. Sediment distributions in pipe flow (Stevens & Charles, 1972) (S_W = weight of sediment/weight of sediment and water).

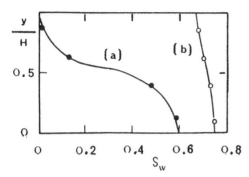

Figure 7.19. Sediment distribution in pipe flow (Charles, 1970).

cross section. For steady flow the diffusion equation for sediment in the lower part of the pipe follows the normal diffusion equation:

$$\varepsilon_s \frac{dS_v}{dy} + \omega S_v = 0 \tag{7.28}$$

For the upper part of the pipe flow, with y' axis taken as positive from the upper boundary towards the pipe center, the diffusion equation can be rewritten as

$$\varepsilon_s \frac{dS_v}{dy'} - \omega S_v = 0 \tag{7.29}$$

The sediment diffusion coefficient is assumed to be proportional to the eddy viscosity and the modified distribution formula of eddy viscosity Equation (7.17) is employed. In an assumption, y_m has been taken equal to the half height of the rectangular closed conduit ($y_m = H/2$). The von Kármán constant κ' in the upper part of the pipe flow is different from κ in the lower part and is assumed to be

$$\kappa' = \alpha\kappa \tag{7.30}$$

Usually α is slightly larger than one. Substituting Equation (7.17) into Equation (7.28) and (7.29), transforming the axis y' into y by $y' = 2y_m - y$, integrating the equations and determining the integration constants from continuity of concentration, the following concentration distribution is obtained

$$\frac{S_v}{S_{vb}} = \begin{cases} \left(\dfrac{(y_m - y)}{y}\right)^Z, & 0 < y \le ay_m \\[2ex] \left(\dfrac{(1 - a)}{a}\right)^Z \exp\left(\dfrac{-Z}{(1 - a)} \cdot \dfrac{y}{(ay_m - 1)}\right), & ay_m < y \le y_m \\[2ex] \left(\dfrac{1 - a}{a}\right)^Z \exp\left(-\dfrac{Z}{a}\left(1 + \dfrac{\alpha}{(1 - a)} \cdot \left(\dfrac{y}{ay_m} - 1\right)\right)\right), & y_m < y \le (2 - a)y_m \\[2ex] \left(\dfrac{1 - a}{a}\right)^{Z(1 + \alpha)} \exp\left(-\dfrac{1 + \alpha}{a}Z\right)\left(\dfrac{y - y_m}{2y_m - y}\right)^{-(\alpha Z)}, & (2 - a)y_m < y < H \end{cases} \tag{7.31}$$

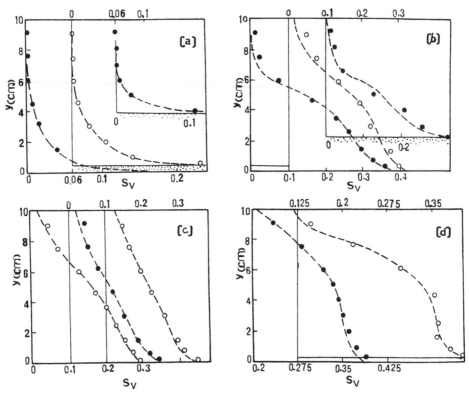

Figure 7.20 Comparison of Equation (7.31) with measured concentration profiles in hyperconcentrated pipe flow.

where S_{vb} is the reference concentration measured at $y = 0.5\, y_m$.

In the experiment by Wang & Qian (1984), the group fall velocity of the suspended load followed the formula

$$\omega = \omega_0(1 - S_v)^7 \tag{7.32}$$

where $\omega_0 = 1.89$ cm/s. Figure 7.20 gives a comparison of the measured concentration distributions and the theoretical curves based on Equation (7.31), with κ and α determined from the measured velocity distributions (see section 7.2.2). In a wide range of average concentration the agreement between the formula and the measured data is quite satisfactory. At low average velocities and concentrations the measured and calculated data show distributions similar to those observed in open channel flow (Figure 7.20a). With increasing concentrations a turning point appears in the central zone and the distribution curve is convex upward in the lower part and concave downward in the upper part (Figure 7.20b). With a further increase in velocity the distribution curves tend to follow a linear relationship (Figure 7.20c). At very high concentrations the gradient of the concentration distribution curve approaches zero in the lower part of the flow (Figure 7.20d). If the effect of lift force due to velocity gradient (Magnus effect) is taken into account, a positive concentration gradient in the lower zone is possible.

7.2.4 *Friction head loss of hyperconcentrated two-phase flow in closed conduit*

The friction head loss of hyperconcentrated two-phase flow in closed conduit is of great importance for hydrotransport. It has been receiving great attention from scientists and engineers throughout the world. Since the First International Conference on the Hydraulic Transport of Solids in Pipes was held in 1970, the study on this topic has been greatly promoted. Durand (1953) initiated his analysis by plotting the pressure gradient J_m as a function of the average velocity U with volume concentration S_v as a parameter. Almost invariable investigations to date have followed Durand and attempted to determine the excess friction head loss $J_m - J_w$ in dimensionless form as a function of the average velocity, in which J_w is the pressure gradient of flow of clear water. This monograph will not discuss these results in detail but give a brief idea of the mechanism of the friction loss.

It is often possible to identify four flow patterns: (1) stationary bed and saltation; (2) sliding bed and heterogeneous suspension; (3) heterogeneous suspension; and (4) pseudo-homogeneous suspension. A typical relationship between the pressure gradient and the average velocity is shown in Figure 7.21 (Vocaldo & Charles, 1972). The head loss of slurry flow is much greater than that of pure liquid flow if the velocity is low and is close to that of pure liquid flow if the velocity is high. Figure 7.22 shows empirical relationship between the head loss of hyperconcentrated two-phase flow in a closed conduit (18×10 cm^2) and

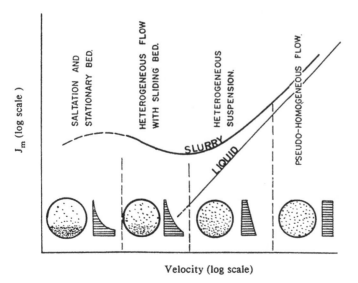

Figure 7.21. Typical relationship between pressure gradient and average velocity for slurries flowing in horizontal pipes.

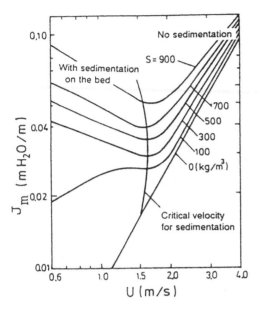

Figure 7.22. Measured relationship between J_m and U and S (Z. Wang, 1984).

the average velocity for sediment (D_{50} = 0.15 mm) concentration from 100 to 900 kg/m^3 (Wang & Qian,1984). There is a minimum point at about U = 1.8 m/s in the J_m – U curves, at which most of the particles are suspended. The point is important because hydrotransport under such a condition is the most economical from the viewpoint of energy consumption. At lower velocity the turbulence intensity is lower, much more energy is needed to transport the same volume of sediment because quite a portion of sediment moves as bed load. The lower the velocity the greater the pressure gradient. We have discussed in Chapter 2 that suspended load consumes only turbulent energy and has little influence on the head loss. With velocity higher than that of the minimum point, the head loss of hyperconcentrated two-phase flows approaches to that of clear water flow.

7.2.5 *Comparison between pipe flow and open channel flow*

Hyperconcentrated flows in pipes are usually with much higher velocity than those in open channel flow, but those in open channels are often much greater in scale. They share mechanism in many aspects. Using two kinds of sediment with diameters 0.2 mm and 5.5 mm, Shen & Wang (1970) investigated incipient motion of sediment in pipe flow and found that Shield's criterion is applicable to flows in rectangular closed conduit. Figure 7.23 shows the results of the experiment which agree Shield's diagram well. Ripples formed on the sediment bed in the closed conduit if the 0.2 mm sediment was used and did not occur if the 5.5 mm sediment was used. The situation was the same as that would be expected in open channel flow.

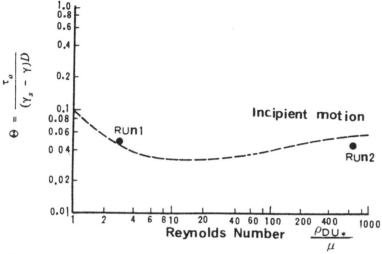

Figure 7.23. Comparison of incipient motion of sediment in pipe flow with Shield's diagram (Shen & Wang, 1970).

Wilson (1985) studied the threshold velocity of suspension of sediment particles in pipe flow and presented the following formula:

$$U_s = 0.6 \, \omega_o \sqrt{\frac{8}{f}} \, e^{0.45 D/d} \qquad (7.33)$$

where D and d are the particles' diameter and the pipe diameter, respectively, f is the Darcy-Weisbach friction factor, ω_o is fall velocity of single particle in water. If $D/d < 0.01$, $e^{0.45 \, D/d} \doteq 1$. By employing $8/f = (U/U_*)^2$, Equation (7.33) is transformed into

$$\frac{\omega_o}{\kappa U_*} = Z = 4.17 \qquad (7.34)$$

This criterion is about the same as that in open channel flow. Bagnold suggested $Z = 3$ as the criterion for sediment suspension in open channel flow, Qian & Wan (1983) employed $Z = 5$ as the criterion, and we suggested $Z < 3$ for suspended load, $Z > 5$ for bed load, and transition from bed load to suspended load if Z is between the two values (see Chapter 2).

If there is stationary sediment on the bed in a steady flow in horizontal closed conduit, the flow transports sediment at its carrying capacity, and the average sediment concentration S_* is used to represent the sediment-transporting capacity. Many formulas for sediment-transporting capacity developed for open channel flow can not be used in pipe flow because the average concentration in pipe flow can vary for more than 100 times at the same average velocity. Such a situation is

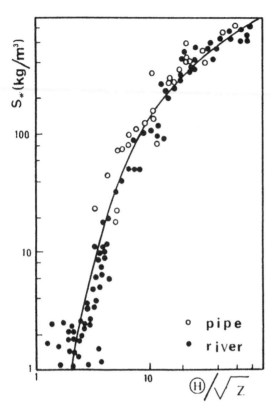

Figure 7.24. Comprehensive correlation of sediment transporting capacity of flows in pipelines and rivers.

impossible in open channel flow. In Figure 7.22 the average sediment concentration varies from 100 to 900 kg/m^3 at the same velocity in the region with sedimentation on the bed. Using $\Theta/\sqrt{Z}$ as a synthetic parameters, Wang and Dittrich (1992) found that the measured data of sediment-transporting capacity in river flows and pipe flows can be unified, as shown in Figure 7.24, in which the parameter Θ is defined by:

$$\Theta = \frac{\tau_o}{(\gamma_s - \gamma)D} \tag{7.35}$$

where τ_o is the shear stress on the bed. In the figure only the data with $Z > 0.1$ (bed material load) and in flows over sediment bed were plotted. Data from pipe flows and rivers follow the same trend. In pipe flow with bed load motion and sediment bed, not only an increase in turbulence intensity involves enhancement of sediment concentraion, but also an increase in the pressure gradient or the parameter Θ enhances bed load motion and raises concentration near the bed, and consequently increases the average concentration. Therefore, the sediment transporting capacity S_* can also be high if the flow is at low velocity but the head loss is quite high in this case.

7.3 LAMINATED LOAD MOTION

As early as in 1879, P. Duboys proposed a model of sediment motion in which bed load particles move in layers under shear. This model, however, has been paid little attention owing to its unconformity with observation of ordinary bed load motion. The development of hyperconcentrated flow in recent years has revived the model. Through experiments with cohesionless particles and water, Bagnold (1955) found that turbulence may vanish and the two-phase flow become a laminar one. In such a flow the effective weight of particles is supported by the dispersive force.

Laminated load motion can also take place in an open channel bed when the main flow is turbulent. Qian and Wang observed in a flume experiment with plastic sand as the solid phase that following the increase in shear stress grain motion penetrated into the bed. Since the grains in the bed were confined by the surrounding particles they could move only in layers and thereby named the laminated load (Qian & Wan, 1983).

Employing the concept of dispersive force, Takahashi (1978) studied water-rock debris flow and developed a velocity distribution formula. Scientists in the United States and Canada carried out a great deal of research in granular flow which is essentially the laminated load motion. With a micromechanical view-point transport coefficients were obtained by statistically averaging the dynamic processes that occurred at the individual particle level. The research was motivated by attempts to better model bin and hopper flows (Shen, 1985; Savage, 1979).

Wilson et al. (1972), Banting & Street (1970) studied the laminated load motion in pipelines they called 'dense phase flow'.

Inspired by Bagnold's work, Wang & Qian (1985, 1987) studied the mechanism of laminated load motion in developing stage and developed stage and tried to apply the theory into debris flow (Qian & Wang, 1984; Wang & Qian, 1989). In developing stage, laminated load motion occurs only in a part of the flow, and the flow may be turbulent, and concentration of laminated load distributes uniuni-formly. In developed stage, however, the flow is laminar and concentration distributes uniformly.

7.3.1 *Dispersive stresses*

The mechanism of the dispersive force has been discussed and the Bagnold's formulas has been presented in Chapter 2. Figure 7.25 shows the results of Bagnold's experiment in which T and P are the dispersive shear stress and dispersive pressure.

Wang & Qian (1985) developed a new formula of dispersive force in laminated load motion. Assume each particle in a layer collides m times in unit time with particles in the lower layer as an average. For each collision, a particle changes its

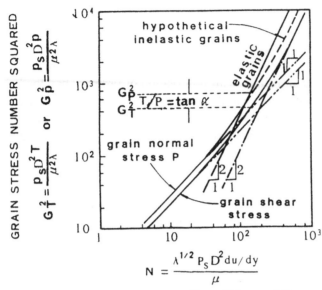

Figure 7.25. Bagnold's results for grain stresses.

momentum by $\rho_s D^3 \Delta \vec{V} \pi / 6$. The dispersive force between the two layers is then

$$F = \frac{\pi \rho_s D^3 m \Delta \vec{V}}{6} \frac{1}{(bD)^2} \qquad (7.36)$$

where $\Delta \vec{V}$ is the average of velocity change after collision, bD the distance between the center of two neighboring particles. The instantaneous velocity of a particle is written as

$$\vec{V} = (u + u')\vec{i} + v'\vec{j} \qquad (7.37)$$

where u is the mean velocity of the particles' layer. By assuming the fluctuating velocity components u' and v' normal random variables, the probability density distribution of v' is written as

$$\phi(v') = \frac{1}{2\pi\sigma} \exp\left(-\frac{v'^2}{2\sigma^2}\right) \qquad (7.38)$$

where σ is the root mean square of v' and has a dimension of velocity. The average distance between particles equals D/λ, where λ is the linear concentration defined by Equation (2.5). The following formula is obtained

$$m = \frac{\lambda}{2kD} \int_{-\infty}^{+\infty} |v'| \phi(v') dv' = \frac{\lambda\sigma}{k\sqrt{2\pi}D} \qquad (7.39)$$

Analyzing the momentum exchange during collisions and substituting the expression of $\vec{\Delta V}$ and Equation (7.39) into Equation (7.36), Wang & Qian (1985) obtained:

$$T = -k'f(\lambda)\rho_s y_m \left(\frac{\omega_o}{U_*}\right)^2 u \frac{du}{dy} \tag{7.40}$$

where T is the dispersive shear stress, y_m is the distance of the point of the maximum velocity of the laminated load from the bed, $f(\lambda)$ is defined by

$$f(\lambda) = \frac{\sqrt{2\pi}}{12} \frac{\lambda^2}{1 + \lambda} \left[1 - \frac{1}{12}\left(1 + \frac{1}{\lambda}\right)^2\right] \tag{7.41}$$

The main difference between Equation (7.40) and the dispersive shear stress formula advanced by Bagnold (Equation (2.4)), is T proportional to $u(du/dy)$ rather than to $(du/dy)^2$. Bagnold abstracted his formula from an experiment in a set-up similar to a rotating-coaxial-cylinder viscometer. It is proved that the results from Bagnold's experiments can also be expressed in the similar form as Equation (7.40) (Wang & Qian, 1987).

In the study on granular flow, many new formulas for the dispersive stress have been advanced. Shen & Achermann (1982) explicitly quantified the stress induced by particle collisions without gross empirical coefficients. Using basic mechanic laws, they studied inelastic and frictional binary collisions. It was assumed that the collisions were generated by random motion of particles. Statistical averaging was performed over all possible geometric arrangements between colliding particles. Through an energy balance argument the intensity of the random motion was shown to depend on the material properties and the mean flow gradient. The stresses in a shear flow of spheres are presented as follows:

$$T = \frac{1}{2} S_{vm}\rho_s D^2 \lambda \left[\frac{(1 + \xi)^3 (0.053 + 0.081\mu)^3 \left(\frac{\lambda}{1 + \lambda^3}\right)^3}{\frac{3}{2}\frac{C_D}{\lambda}\frac{\rho}{\rho_s} + \frac{1 + \xi^2}{4} + \frac{\mu(1 + \xi)}{\pi} - \frac{\mu^2(1 + \xi)^2}{4}} \right]^{0.5} \left(\frac{du}{dy}\right)^2 \tag{7.42}$$

where ξ is sphere's restitution coefficient, and μ the viscosity of the liquid.

The effects of friction and inelasticity in rapid flow of granular materials have also discussed by Cook & Franklin (1964, 1971), Brahic (1975, 1977), Goldreich & Tremaine (1978a, 1978b), Shahinpoor & Siah (1981), Ahmadi & Shahinpoor (1982, 1983). They in particular offered the following expression for the dispersive shear stress and dispersive pressure in a steady Couette flow

$$T = \frac{\pi}{6} \frac{k^2\rho_s f\lambda D^2}{1 + (1 + \beta)\mu_k^2} (\cos\alpha + \mu_k \sin\alpha) [(1 + \beta)\mu_k \sin\alpha + \cos\alpha] \left(\frac{du}{dy}\right)^2 \tag{7.43}$$

$$P = \left(\frac{\mu_k \cos\alpha - \sin\alpha}{\cos\alpha + \mu_k \sin\alpha}\right) T \tag{7.44}$$

where k, α and β are material parameters, μ_k is a kinematic friction factor, and f is the frequency of collision and is a function of λ.

Equations (7.42) and (7.43) agree with Bagnold's formula in T proportional to $(du/dy)^2$. But Equation (7.40) gives the dispersive shear stress proportional to $u(du/dy)$. The difference resulted from mathematical manipulation on fluctuating velocity of grains. In the former the fluctuating velocity was assumed proportional to velocity gradient du/dy, and in the latter it was assumed proportional to mean velocity u. Compbell & Brennen (1983) modelled a granular Couette flow and got the fluctuating velocities in the three directions. Drake made an experiment and measured distribution of granular temperature (velocity fluctuation of particles). Both studies revealed the fluctuating velocity of particles proportional to the mean velocity rather than velocity gradient. It seems that Equation (7.40) is more reasonable.

7.3.2 *Velocity and concentration distributions of laminated load in developing stage*

Takahashi (1978) studied the velocity distribution of coarse particles through experiment in a flume. A sand bed was molded in the flume and set a high angle of inclination. The bed was saturated with seepage water at first. A discharge of water was then suddenly supplied at the upstream end, making the mixture of grains and water start moving rapidly downstream. The driving shear stress is

$$\tau = S_v(\rho_s - \rho)g\,(H - y)\sin\theta \tag{7.45}$$

where θ is the slope angle. If the flow is steady, the driving stress must equal the dispersive shear stress. Employing Bagnold's formula he established

$$a_i\rho_s\,\lambda^2 D^2\left(\frac{du}{dy}\right)^2\cos\alpha = S_v(\rho_s - \rho)g\,(H - y)\sin\theta \tag{7.46}$$

Integration of the equation under the boundary condition $u = 0$ at $y = 0$ yields

$$u = \frac{2}{3}D\left\{\frac{g\sin\theta}{a_i\sin\alpha}\left[S_v + (1 - S_v) - \frac{\rho}{\rho_s}\right]\right\}^{1/2}\left[\left(\frac{S_{vm}}{S_v}\right)^{1/3} - 1\right]$$

$$\left[H^{3/2} - (H - y)^{3/2}\right] \tag{7.47}$$

where H is the depth of the flow. By denoting u_m the maximum velocity at the surface of the flow, Equation (7.47) is rewritten as

$$\frac{u}{u_m} = 1 - \left(1 - \frac{y}{H}\right)^{3/2} \tag{7.48}$$

In the derivation the concentration was assumed uniformly distributed along the

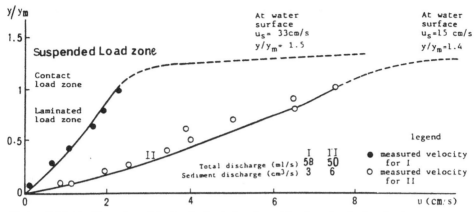

Figure 7.26. Profiles of particles' velocity in laminated load motion in developing stage.

vertical. It is questionable because the solid concentration does vary with y. On the other hand the velocity distribution formula agrees with the measured data.

From their experiments Wang & Qian (1987) found that laminated load motion in open channel is generally in developing stage with heterogeneous concentration distribution, the velocity distribution is similar to that given by Equation (7.48).

Figure 7.26 shows the velocity distribution of particles of laminated load motion in developing stage. The flow is turbulent and there are a few suspended particles in the upper part and the laminated load motion takes place in the lower zone, $y < y_m$. The velocity increases almost linearly from $y = 0$ to $y = y_m$ and rises sharply in the upper flow.

The concentration distribution of particles must be known if a velocity formula is developed from the dispersive stress formulas, because the dispersive shear stress depends on the linear concentration. Wang & Qian (1987) assumed the following concentration distribution of laminated load in developing stage:

$$\frac{\lambda^2}{1 + \lambda} = \xi \frac{1 - \dfrac{y}{y_m}}{\sqrt{\dfrac{y}{y_m}}} \tag{7.49}$$

or

$$\frac{\left(\dfrac{S_v}{S_{vm}}\right)^{2/3}}{1 - \left(\dfrac{S_v}{S_{vm}}\right)^{1/3}} = \xi \frac{1 - \dfrac{y}{y_m}}{\sqrt{\dfrac{y}{y_m}}} \tag{7.50}$$

where ξ is a coefficient and equals the linear concentration at $y = 0.38\, y_m$. Figure 7.27b shows the curves of Equation (7.49) or (7.50) for $\xi = 10$ and $\xi = 50$.

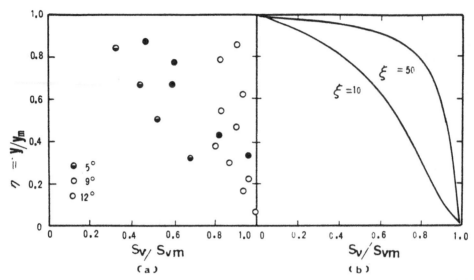

Figure 7.27. Concentration distributions in laminated load zone: a) Measured by Tsubaki et al. (1983); b) Postulated by Wang & Qian (1987).

Table 7.1. Main parameters of Wang's experiments (L. Wang, 1986).

Run	Whole flow depth H (cm)	Relative thickness of laminated load zone y_m/H	Average velocity U (cm/s)	Average concentration S_v/S_{vm}	Slope J (%)	Linear concentration at $y = 0.38\,y_m$ ξ
1	3.1	0.8	72.3	0.635	6.25	18
2	3.0	0.8	77.4	0.633	5.70	18
3	3.7	1.0	71.8	0.529	5.66	18
4	3.2	1.0	69.1	0.537	6.15	18
5	3.0	1.0	58.8	0.541	6.27	18
6	3.4	1.0	68.3	0.376	4.25	18
7	3.4	1.0	71.5	0.339	5.07	18

Tsubaki et al. (1983) measured the concentration distribution of laminated load in a flume experiment. Figure 7.27a shows the measured results in three runs with bed slope 5°, 9° and 12°. The measured data conform with the curves in Figure 7.27b in principle. Wang (1986) checked the formulas (7.49) and (7.50) with measured data. He did an experiment in a flume with a plastic sand of diameter 0.47 mm and specific gravity 1.227. Laminated load motion developed if the slope of the flume was high enough. He measured the concentration distribution by using a radioisotope transmission gauge. Table 7.1 shows the main parameters of the experiment. Figure 7.28 presents a comparison of the measured concentration distribution of laminated load with Equations (7.49) or (7.50). The results proved that the concentration distribution of laminated load well follows the curves given by Equation (7.49) or (7.50).

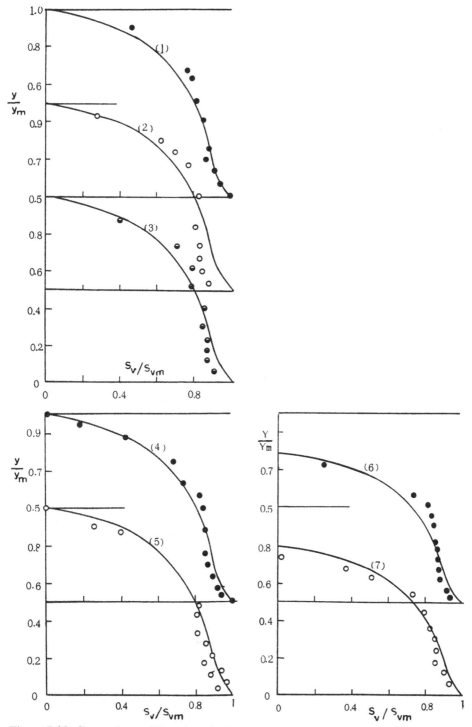

Figure 7.28. Comparison of Equation (7.50) with measured concentration profiles in laminated load motion (L. Wang, 1986).

Employing the concentration distribution formula (7.49) and the expression of dispersive shear stress Equation (7.40), Wang & Qian (1987) obtained the following equation:

$$\tau_0 \left(1 - \frac{y}{y_m}\right) = kA\rho_s y_m \left(\frac{\omega_o}{U_*}\right)^2 \frac{1 - \frac{y}{y_m}}{\sqrt{\frac{y}{y_m}}} u \frac{du}{dy} \tag{7.51}$$

where

$$A = \xi \frac{\sqrt{2\pi}}{12} \left[1 - \frac{1}{12}\left(1 + \frac{1}{\lambda}\right)^2\right] \tag{7.52}$$

The value of A varies only 26% in the range $1 \leq \lambda < \infty$. It can be regarded as a constant for all practical purposes. Integration of Equation (7.51) yields

$$\frac{u}{u_m} = \left(\frac{y}{y_m}\right)^{3/4} \tag{7.53}$$

where u_m is the velocity at $y = y_m$. Figure 7.26 shows the curves of the velocity distribution formula which follow the measured data well.

Equation (7.53) and Takahashi-Bagnold formula Equation (7.48) and a velocity distribution formula developed by Tsubaki et al. (1983) are plotted in Figure 7.29 for comparison. The formula developed by Tsubaki et al. is very complicated and we do not discuss it in detail here. The data measured by Wang & Qian, Takahashi, and Tsubaki et al. are also plotted in the figure and compared with the formulas. The agreement between the formulas and measured data is satisfactory.

Lots of researchers found that laminated load motion in developing stage can be subdivided into two zones: a viscous zone and an inertia zone. The viscous zone is thin and near the bed in which the velocity distributes differently from the inertia zone. The distribution curve in the viscous zone is convex and in the inertia zone is concave. The data and the distribution curve presented by Tsubaki et al. (1983) in Figure 7.29 shows that about 30% of the flow is in viscous zone.

Assuming the concentration distribution follows

$$\frac{S_v}{S_{vm}} = 1 - c\frac{y}{y_m} \tag{7.54}$$

and employing the dispersive shear stress formula Equation (7.40), Wang (1986) derived the following velocity distribution formula in viscous zone:

$$\frac{u}{u_1} = \left(\frac{y}{y_1}\right)^{3/2} \tag{7.55}$$

where y_1 is the distance from the bed of the point dividing the viscous zone and inertia zone, and u_1 is the velocity at $y = y_1$. For the inertia zone, he developed the following formula:

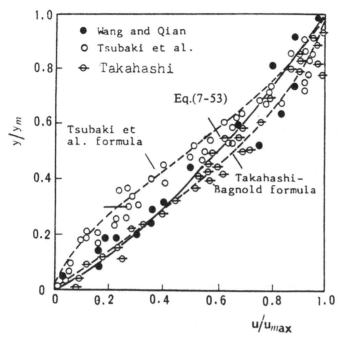

Figure 7.29. Velocity profiles of laminated load suggested and measured by different researchers.

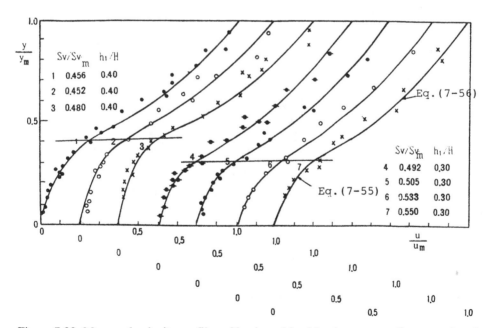

Figure 7.30. Measured velocity profiles of laminated load in viscous zone (lower part) and inertial zone (upper part), compared with Equations (7.55) and (7.56).

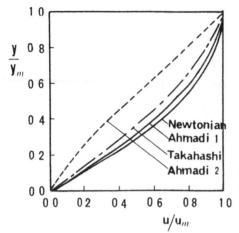

Figure 7.31. Comparsion of velocity profiles of granular flow.

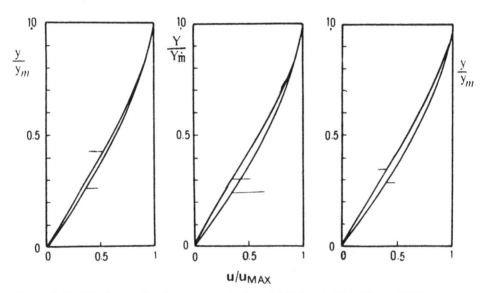

Figure 7.32. Velocity profiles from a numerical model (Schweiwill & Hutter, 1983).

$$\left(\frac{u}{u_1}\right)^2 = 1 + \frac{y^{3/2} - y_1^{3/2}}{y_m^{3/2} - y_1^{3/2}}\left(1 + \left(\frac{u_m}{u_1}\right)^2\right) \tag{7.56}$$

If the whole laminated load motion is in the inertia zone, $h_1 = 0$ and $u_1 = 0$, the formula reduces to

$$\frac{u}{u_m} = \left(\frac{y}{y_m}\right)^{3/4} \tag{7.57}$$

Figure 7.30 shows the measured velocity distribution and the curves of Equations

(7.55) and (7.56) (Wang, 1986). It seems that the equations agree with measured data quite well.

Ahmadi (1983) studied flow of granular materials and developed two velocity profiles, as shown in Figure 7.31. One profile is close to the Takahashi-Bagnold formula and another one close to the Tsubaki et al.'s formula.

Schweiwiller & Hutter (1983) analysed flow of granular material down a chute and obtained a numerical solution. The velocity profile from the numerical solution is shown in Figure 7.32. The distribution is close to Wang and Qian's formula.

7.3.3 *Velocity distribution of laminated load in fully developed stage*

If the energy supply is high enough the laminated load motion can develop in the whole flow field and there is no sand bed and suspended load zone. Such a fully developed laminated load motion usually occurs in closed conduit. In fully developed stage the concentration of laminated load distributes uniformly but the velocity distribution is quite different from the velocity distribution formulas presented above. Wang & Qian (1985) made an experiment with plastic sand of diameter $D = 0.25$ mm and specific gravity 1.057, in a closed conduit with cross section 18×10 cm^2. With low concentration of particles the flow is turbulent and the velocity distribution followed the logarithmic formula as shown in Figure 7.33 (dash line). Following increase in concentration and decrease in average velocity, turbulent eddies were suppressed by the dispersive stress and the flow transformed to the laminated load motion, the velocity distribution was closer and closer to the solid line curves that is given by Equation (7.58). Observations proved that turbulent eddies of scale larger than the particle's size disappear altogether and the solid particles move in layers if the concentration is sufficiently high ($S_v > 50\%$). The concentration of particles very uniformly distributed in the whole flow field. Employing Equation (7.40) and assuming the dispersive shear stress T is equal to the tractive shear stress $\tau = \tau_o (1 - y / y_m)$ in steady flow, Wang

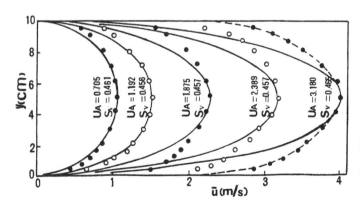

Figure 7.33. Velocity profiles from suspended load motion to laminated load motion.

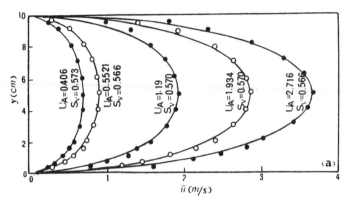

(a) Pipe: H=10 cm, B=18 cm

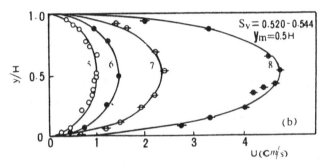

(b) Pipe: H=3 cm, B=3 cm

Figure 7.34. Velocity profiles of laminated load motion in fully developed stage: a) Plastic sand: $D = 0.25$ mm, $\gamma_s = 1.057$ g/cm^3; b) Polyvinyl chloride grains $D_{50} = 0.68$ mm, $\gamma_s = 1.295$ g/cm^3.

& Qian (1985) obtained the following formula of velocity distribution:

$$\frac{u}{u_m} = \left[1 - \left(1 - \frac{y}{y_m}\right)^2\right]^{\frac{1}{2}} \tag{7.58}$$

Figure 7.34a shows the measured velocity distributions and curves of the velocity distribution formula. The measured data follow the curves quite well. The velocity distribution was further checked with measured data in an experiment with polyvinyl chloride grains, of median diameter $D_{50} = 0.68$ mm and specific gravity 1.295, in a square pipeloop of cross section 3×3 cm^2. Figure 7.34b shows the measured velocity distributions and comparison with Equation (7.58) (Wang & Qian, 1987). The result provides further proof of the velocity distribution of Equation (7.58).

Figure 7.35 provides a comparison of the velocity distributions of laminated load motion in developing stage and developed stage, of turbulent flow and

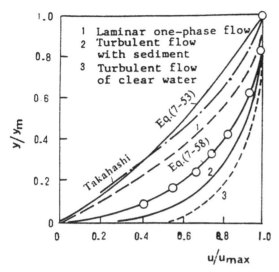

Figure 7.35. A comparison of various velocity profiles.

laminar flow of Newtonian fluid, and of turbulent flow carrying sediment. The velocity distribution of fully developed laminated load motion falls somewhat between that of turbulent flow and laminar flow, and the distribution of laminated load in developing stage is close to that of laminar flow of Newtonian fluid but the former has more uniform velocity gradient. We define

$$C_u = \int_0^1 \left[1 - \frac{u}{u_m} \right] d\left(\frac{y}{y_m} \right) \tag{7.59}$$

and call C_u the center of the velocity distribution, where u_m is the maximum velocity in the profile. The value of C_u is 0.43 for laminated load motion in deveping stage (Equation (7.53)), and 0.40 for Takahashi's formula. It equals 0.333 for laminar flow of Newtonian fluid and equals 0.215 for laminated load motion in developed stage (Equation (7.58). Turbulent flow has the smallest C_u value. If the logarithmic velocity distribution formula is used, one can obtain $C_u = U_*/ \kappa u_m$. Because U_*/ u_m is usually in the range of 0.06-0.02, the value of C_u is in the range of 0.15-0.05. Velocity profile with large C_u value is not stable, therefore, there is no velocity profile with C_u value larger than 0.5.

7.3.4 *Relative velocity between particles and fluid*

Laminated load is a kind of bed load and the motion consumes energy supplied by the fluid phase. Therefore, the velocity of fluid must be greater than that of particles. Wang et al. (1990) conducted an experiment in an open channel with water and clay suspension as the liquid phase and with gravels of median diameter $D_{50} = 10$ mm and specific gravity 2.65 as the solid phase. The velocities of grains

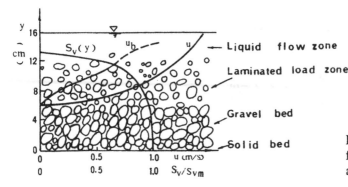

Figure 7.36. Velocity profiles of grains and liquid and concentration profile.

and the liquid were quite different. Figure 7.36 shows the velocity distribution of grains and that of the liquid, the relative velocity (the velocity of the liquid u minus the velocity of the grains u_b) is rather great.

By the relative velocity $u - u_b$, the liquid phase acts a drag force F on the particles

$$F = C_D \frac{\pi D^2}{4} \frac{\rho (u - u_b)^2}{2} \tag{7.60}$$

where C_D is the drag factor, which is a function of the Reynolds number $Re = D(u - u_b) / \nu$. Let $n = 6S_v / \pi D^3$ be the number of the particles in unit volume, the energy delivered to the particles in unit time is

$$e_b = nFu_b = \frac{3}{4} C_D \frac{S_v \rho u_b (u - u_b)^2}{D} \tag{7.61}$$

For a steady laminated load motion in an open channel, the following energy equation can be established:

$$\gamma_s S_v u_b \tan \theta + e_b = T \frac{du_b}{dy} - \frac{d(Tu_b)}{dy} \tag{7.62}$$

The dispersive shear stress is proportional to the total submerged weight of laminated load above the elevation y

$$T = \int_y^{y_m} (\gamma_s - \gamma) S_v \tan \alpha \, dy \tag{7.63}$$

Substituting Equations (7.63) and (7.61) into Equation (7.62), one obtains

$$\frac{(u - u_b)^2}{gD} = \frac{4}{3} \frac{1}{C_D} \left[\left(\frac{\gamma_s}{\gamma} - 1 \right) \tan \alpha - \frac{\gamma_s}{\gamma} \tan \theta \right] \tag{7.64}$$

where $\tan \theta$ is the slope of the channel, $\tan \alpha$ is the dynamic frictional coefficient. If the slope is small, main energy comes from the liquid phase and the relative

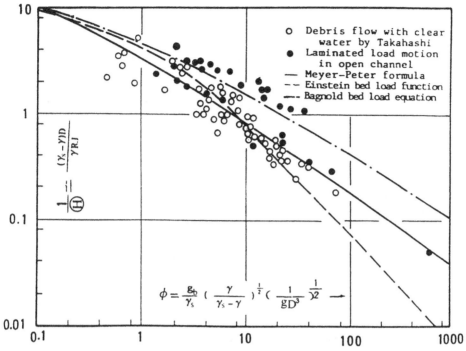

Figure 7.37. Measured rate of laminated load transport compared with bed load formulas.

velocity is great. The drag factor C_D increases with increasing concentration and therefore the relative velocity is small in the lower flow where the concentration is high.

7.3.5 *Sediment transporting capacity in laminated load motion*

Because the laminated load motion is a kind of bed load motion, the rate of sediment transport is comparable to that of bed load motion. Qian (1980) demonstrated that most bed load formulas can be transformed into a relationship between the dimensionless shear stress Θ and the bed load transport intensity Φ, as curves in Figure 7.37. The data of laminated load motion measured by Takahashi (1978), Wang & Qian (1987), and Wang et al. (1990) are plotted in this way in Figure 7.37. The points distribute around the Meyer-Peter formula and Bagnold formula.

REFERENCES

Ahmadi, G. & M. Shahinpoor 1982. New evolutionary equations for fluctuation in rapid granular flow. *Proc. Abs. 19th SESI Conf., Rolla, Mo.* p.256.

Ahmadi, G. & M. Shahinpoor 1983. On collision operators in rapid granular flow. *Power Tech., J.*

Ayukawa, K. 1979. Velocity distribution and pressure drop of heterogeneously suspended flow in hydraulic transport through a horizontal pipe. *Hydrotransport 1, First International Conference on the Hydraulic Transport of Solids in Pipes*, pp. F3-33-43.

Bagnold, R.A. 1954. Experiments in a gravity-free dispersion of large spheres in a Newtonian fluid under shear. *Proc. Royal Soc., London, Ser. A*, Vol. 225: 49-62.

Bagnold, R.A. 1955. Some flume experiments on large grains but little denser than the transporting fluid, and their implications. *Proc. of Inst. of Civil Engrs.*, 4(3): 174-205.

Bagnold, R.A. 1956. *Flow of cohesionless grains in fluid*. Philos. Trans., Royal Soc. of London, Ser. B, 249: 235-297.

Banting, R.A. & M. Streat 1970. Dense-phase flow of solids mixtures in pipelines. *Proc. of Hydrotransport* 1, Paper C1.

Brahic Andre 1975. A numerical study of a gravitating system of colliding particles: Application to the dynamics of Saturn's rings and to the formation of solar system. *Icarus*, Vol. 25: 452-458.

Brahic Andre 1977. System of colliding bodies in a gravitational field: I-numerical simulation of the standard model. *Anstron. Anstrophys.*, Vol. 54: 895-907.

Campbell, C.S. & C.E. Brennen 1983. Computer simulation of shear flow of granular materials. *Mechanics of granular materials, new models and constitutive relations*, Jenkino, J.T. and Satake, Elsevier.

Charles, M.E. 1970. Transport of solids by pipeline. *Proc. of Hydrotransport 1, BHRA, Cranfield*, Paper A3.

Chien, N. (Ning Chien) & Z. Wan 1965. The effects of sediment concentration gradient on the characteristics of flow and sediment motion (in Chinese). *J. of Hydraulic Engineering*, No. 4: 1-20.

Cook, A.F. & F.A. Frankling 1964. Rediscussion of Maxwell's Adams Prize essay on the stability of Saturn's Rings'. *Astron, J.*, Vol. 69(2): 173-200.

Einstein, H, A. & N. Chien 1955. *Effects of sediment concentration near the bed on the velocity and sediment distribution*. M.R.D. Sediment Series No.8, Missouri River Div., Corps of Engrs.

Frankling, F.A., A.F. Cook & G. Colombo 1971. A dynamical model for the redial structure of Saturn's rings' II. *Icarus*, 15: 80-92.

Friedman, G.M. & J.E. Sanders 1978. *Principles of sedimentology*. John Wiley & Sons, Inc., pp.100.

Goldreich, P. & S. Tremine 1978a. The velocity dispersion in Saturn's rings'.*Icarus*, Vol. 34: 240-253.

Goldreich, P. & S. Tremine 1978b. The formation of the Cassini division in saturn's rings'. *Icarus*, Vol. 34: 240-253.

Govier, G.W. & K. Aziz 1972. *The flow of complex mixture in pipes*. Van Nostrand Reinhold Co., pp 792.

Jobson, H.E. & W.W. Sayre 1969. An experimental investigation of the vertical mass transfer of suspended sediment (a). *Thirteenth congress of the International Association for Hydraulic Research, Proceeding*, Vol. 2(B11).

Ismail, H.M. 1952. Turbulent transfer mechanism and suspended sediment in closed channels. *ASCE, transactions*, Vol. 117: 409-434.

Kanatani, K. 1979. A micropolar continuum theory for the flow of granular materials. *Ins. J. Engng. Sci.*, Vol. 17: 419-432.

Katsumasa Yano & Atsuyuki Daido 1969. The effect of bed load movement on the velocity distribution of flow. *Thirteenth congress of the International Association for hydraulic Research, Proceedings*, Vol. 2(B34).

Newitt, D.M., J.F. Richardson & C.A. Shook 1962. Symposium on interaction between fluids and particles. *London. Proc.*, p.87, Published by the Institution of Chemical Engineering, London.

Ogawa, S., A. Omemura & N. Oshima 1980. On the equation of fully fluidized granular material. *ZAMP*, Vol. 31: 483-493.

Qian, N. (Ning Chien) 1980. A comparison of the bed load formulas (in Chinese). *J. Chinese Soc. Hyd. Engin.*, No. 7: 1-11.

Qian, N. (Ning Chien) & Z. Wan 1983. Dynamics of Sediment Movement (in Chinese). *Chinese Science Press, Chapter 9, Beijing.*

Qian, N. & Z. Wang 1984. A preliminary study on the mechanism of debris flow (in Chinese). *Acta Geographica Sinica*, Vol. 39(1): 33-34.

Savage, S.B. 1979. Gravity flow of cohesionless granular materials in chutes and channels. *J. Fluid Mech.*, Vol. 92: 53-96.

Shahinpoor, M. & J.S.S. Siah 1982. New constitutive equations for the rapid flow of granular materials. *J. Non-Newtonian Fluid Mech.*, Vol. 9: 147-156.

Shen, H.H. 1985. An overview of research in the mechanics of a rapidly deforming granular material. *Proceedings of International Workshop of Flow at Hyperconcentrations of Sediment, Beijing*, p. 1-4.

Shen, H.W. & J.-S. Wang 1970. Incipient motion and limiting deposit condition of solid mixtures. *J. Engrg. Mech. Div., ASCE*, Vol. 108(EM5): 748-763.

Silin, N.A., Yu.K. Votoshkin, V.M. Karasik & V.E. Ocheretko 1969. Research on solid liquid flows with high consistency. *Proc. of 13th Congress of IAHR*, 2: 147-156.

Stevens, G.S. & M.E. Charles 1972. The pipeline flow of slurries: Transition velocities. *Proceedings of Hydrotransport*, 2 Paper E3, Organized by SHRA.

Sutton, O.G. 1953. Problems of wind structure near the surface. *Micrometeorology*, pp. 229-272. McGraw-Hill Book Co.

Takahashi, T. 1978. Mechanical characteristics of debris flow. *J. Hyd. Div., Proc., ASCE*, Vol. 104(HY8): 1153-1169.

Tsubaki, T., H. Hashimoto & T. Suetsugi 1983. Interparticle stresses and characteristics of debris flow. *J. Hydroscience and Hydr. Engineering*, Vol. l(2): 67-82.

Vanoni, V.A. 1946. Transportation of suspended sediment by water. *Trans, ASCE*, Vol 111: 67-133.

Vocadlo, J.J. & M.E. Charles 1972. Prediction of pressure gradient for the horizontal turbulent flow of slurries. *Hydrotransport* 2, Paper C1.

Wang, L. 1986. Experimental study on the basic laws of laminated load motion (in Chinese). M.S.Thesis, Department of Hydraulic Engineering Tsinghua University.

Wang, Z. 1984. Experimental study on the mechanism of hyperconcentrated flows (in Chinese). PH.D. Thesis, IWHR, Beijing, China.

Wang, Z. 1986. A study on the mechanism of suspended load motion (in Chinese). *J. of Hydraulic Engineering*, No. 7, pp. 11-20.

Wang, Z. & A. Dittrich 1992. A study on problems in suspended sediment transportation. *Proceedings of the Second International Conference on hydraulic and Environmental Modelling of Coastal, Estuarine and River Waters*, Vol. 2: 467-478.

Wang, Z. & N. Qian (N. Chien) 1984. Experimental study of two-phase turbulent flow with hyperconcentration of coarse particles. *Scientia Sinica*, Series A, Vol. 27(12): 1317-1327.

Wang, Z. & N. Qian (N. Chien) 1985. Experimental study of motion of laminated load. *Scientia Sinica*, Ser, A, Vol. 28(1): 102-112.

Wang, Z. & N. Qian (N. Chien) 1987. Laminated loads: Its development and mechanism of motion, *International Journal of Sediment Research*, No. 1, Nov. 1987, pp. 102-124.

Wang, Z. & N. Qian (N. Chien) 1989. Characteristics and mechanism of debris flow. *Proceedings of 4th Inter. Symposium on River Sedimentation, IRTCES, China Ocean Press*.

Wang, Z., Q. Zeng & X. Zhang 1990. Experimental study of solid-non-Newtonian fluid two-phase flow (in Chinese). *J. of a Sediment Research*, No. 3, pp. 2-13.

Wilson, K.C. 1985. Comparison of hyperconcentrated flows in pipes and open channels, *Proceedings of International Workshop on Flow at Hyperconcentrations of Sediment*, p. 115, IRTCES, Sep., Beijing.

Wilson, K.C., M. Streat & R.A. Banting 1972. Slip-model of correlation of dense two-phase flow, *Proc. of Hydrotransport 2*, Paper B1, BHRA, Cranfield.

YVPO (Yangtze Valley Planning Office) 1985. Report of Preliminary Design of the Three Gorges Project (in Chinese)..

CHAPTER 8

The effect of fine particles on coarse particle movement

8.1 INTRODUCTION

Hyperconcentrated flow in nature often carries various sizes of sediment, including a certain amount of fine particles. In the middle reaches of the Yellow River it is rather common that the suspended sediment becomes coarser and coarser as the concentration increases, while the concentration of fine particles keeps nearly constant. For Huangfuchuan, a tributary of the Yellow River, the concentration of particles finer than 0.007 mm fluctuates around a mean value of 48.5 kg/m^3 for total concentration of 400 kg/m^3 to 1540 kg/m^3 (Qian et al., 1984). The same is true for Wuding River where in the content of solid grains the concentration of grains finer than 0.01 mm is independent of the total concentration and assumes a value of 50 kg/m^3. This seems a common situation for rivers with hyperconcentrations. Similar situation happens in some debris flow. In Xiaojiang River Basin of Yunnan Province, the concentration of particles finer than 0.05 ~ 0.06 mm varies in a narrow range of 300 ~ 500 kg/m^3 while the total concentration varies in a wide range of 500 ~ 2000 kg/m^3.

A certain amount of fine particles in hyperconcentrated flow or debris flow plays an important role and has great effect on the behaviour of the flow.

The existance of fine particles makes fluid viscosity increase and fall velocity of coarse particles reduce. The change of fluid viscosity and fall velocity induces a series of changes in bed configuration, flow resistance, velocity. It seems that the existence of a certain amount of fine particles favours to the transport of coarse particles. On the other hand, too much fine particles result in extremely high viscosity of the mixture and consequent high flow resistance.

Sets of research work on the effect of fine particles have been done. Most of them concentrate on flow resistance, sediment carrying capacity and some engineering problems. Studies on its mechanism and some details have also been started.

The main results of these research works will be reviewed in this chapter.

171

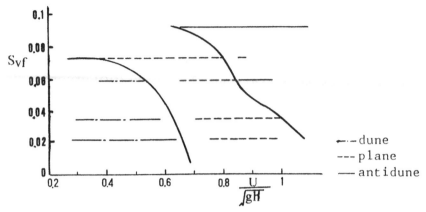

Figure 8.1. The effect of particle concentration S_{vf} on bed configuration.

8.2 BED CONFIGURATION

Wan (1982) carried out a series of experiments in a closed rectangular conduit (30 cm × 20 cm) with a granular bed, which was composed of plastic beads with cylindrical shape. The density and volumetric diameter of plastic beads are $\rho_s = 1.29$ g/cm^3 and $d = 0.345$ cm, respectively. Clear water ran over the granular bed at first, then mud (water-bentonite mixture) at different concentrations ran over the bed successively. By comparing bed configuration moulded by clear water flow with that moulded by flow at different clay concentrations S_{vf}, it was found that the latter had the following characteristics:

1. Particles on the bed started moving at a larger discharge;
2. Bed with dunes transformed to plane at a smaller discharge;
3. As a result, dunes existed in a narrower range of discharge;
4. The dunes were gentler and smoother in shape. The slope of their lee side was no longer the repose angle. They had a more or less symmetric outline.

It was also found that the higher the clay concentration was, the more obvious the differences were. When the clay concentration reached a certain value, dunes did not appear at any discharge.

Wan & Song (1987) conducted some similar experiments in an open channel flume. That is, let clear water at first, then mud with different clay concentrations run over granular bed. The conclusion is similar too, that is, dunes in mud have gentler slope and smoother shape. And they also divided the whole area in $S_{vf} \sim U/\sqrt{gH}$ plot into three different regions, that is, regions of dune, plate, and antidune, see Figure 8.1. It can bee seen from the figure that with S_{vf} increasing dunes disappear and plane bed turns to be antidunes at smaller $U/\sqrt{gH}$. Here U and H are average velocity and depth, respectively.

8.3   PROFILES OF COARSE PARTICLE CONCENTRATION

Song et al. (1986) conducted some experiments in a pipeline with a rectangular cross section of 18 cm × 10 cm. Sand of median diameter $D_{50} = 0.15$ mm was used as coarse particles and Huayuankuo clay, the granulometric composition of which is shown in Figure 8.2, was used as fine particles. Velocity profiles and profiles of coarse particle concentration were measured. Some of the coarse particle concentration profiles are shown in Figure 8.3, in which S_{vc} and S_{vf} are coarse particle concentration, in volume, and concentration of matrix, consisting

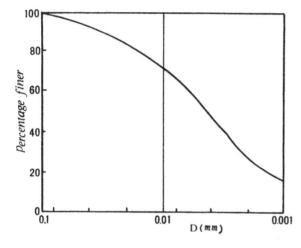

Figure 8.2. Granulametric composition of Huayuankuo clay.

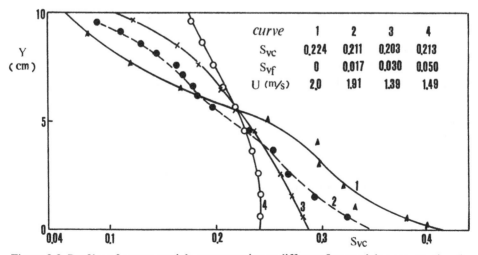

Figure 8.3. Profiles of coarse particle concentration at different fine particle concentration (in conduit).

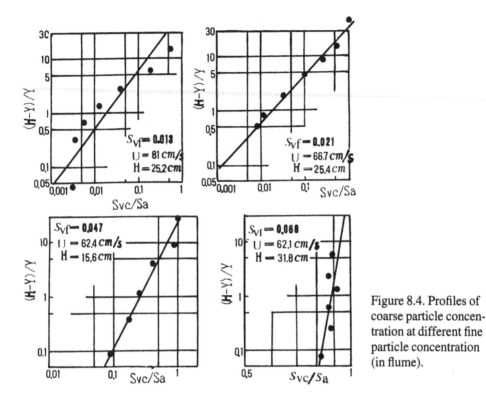

Figure 8.4. Profiles of coarse particle concentration at different fine particle concentration (in flume).

of clay and water, in volume, respectively. y is the location of sampling points counted from the bottom.

Flow with pure suspended coarse particles had the most nonuniform distribution of coarse particle concentration. As clay particles were added and S_{vf} increased step by step, the distribution of coarse particle concentration became more and more uniform.

Profiles of coarse particle concentration were also measured in open channel flume by Wan & Song (1987). Experimental dots distribute nearly along a straight line on log-log plot, see Figure 8.4, in which S_a is the coarse particle concentration at a reference point $y = a$. The slope of the straight line became steeper and steeper with the increase of clay concentration step by step. It means that the diffusion theory of suspended load remains valid in such case and the diffusion index Z, which is the reciprocal of the slope of the straight line, decreases with S_{vf} increasing. According to the diffusion theory,

$$Z = \frac{\omega}{kU_*}$$

here k is the Karman constant, U_* the shear velocity, and ω the fall velocity of coarse particles.

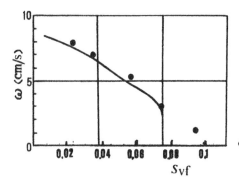

Figure 8.5. The fall velocity of coarse particle at different fine particle concentration.

Based on this equation, fall velocity ω in each set of experiments can be calculated according to the measured k, U_* and Z. On the other hand, fall velocity of coarse particles settling in slurry can also be calculated based on $C_D - Re_2$ relationship, summarized in reference of Qian & Wan (1986) and shown in Figure 5.10.

Fall velocities deduced from diffusion index, which are symbolized by 'O', are compared with fall velocities calculated on the basis of $C_D - Re_2$ relationship, which is represented by a solid line, as shown in Figure 8.5.

For experiments with S_{vf} lower than 0.075 fall velocities obtained from the two different ways agree with each other quite well. At higher clay concentration, according to the force balance of a particle in Bingham fluid, its fall velocity should be zero. But the measured profiles of coarse particle concentration still have a certain gradient. It means ω is not zero in this case but a definite value. The explanation of the contradiction between them can refer to the original paper (Wan & Song, 1987).

Some conclusions can be deduced by summarizing the foregoing facts.

1. The existence of fine particles makes fall velocity of coarse particles reduce. Consequently, the profile of coarse particle concentration is more uniform.

2. The distribution of coarse particle concentration still obeys the law of diffusion theory if the fall velocity of coarse particle has been revised by considering the influence of fine particles.

8.4 VELOCITY PROFILE

In above mentioned experiments fulfilled by Song et al. (1986) velocity profiles were also measured. The comparison between velocity profiles under different conditions is shown in Figure 8.6, in which S_{vc} and S_{vf} are the concentration of coarse particles and that of matrix, respectively, k the Karman constant.

The velocity profile (No. 5) of flow with suspended pure coarse particles deviates from that of the clear water flow (No. 1) the farthest. As clay particles

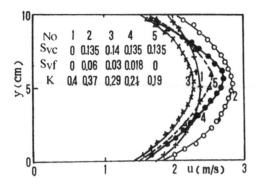

Figure 8.6. Velocity profiles at different fine particle concentration.

were added step by step, the velocity profiles approached that of the clear water flow gradually. As an index reflecting the uniformity of velocity distribution, Karman constant k increased from 0.19 to 0.37, correspondingly.

The explanation of such variation is as follows. The existence of clay particles made fluid viscosity increase and the fall velocity of coarse particles reduce. Consequently, profiles of coarse particle concentration became more uniform and the density gradient formed by coarse particles reduced, as we discussed before. Therefore, energy coming from turbulence for suspending coarse particles reduced and velocity profiles tended to be more uniform.

Song (1986) did some analysis on the velocity profile along the centerline of a pipeline with a rectangular cross section. Improving Ismail's hypothesis on momentum exchange coefficient ε_m, Song used three parabolic segments to replace the constant. For clear water flow the momentum exchange coefficient ε_m is as follows:

$$\varepsilon_m = \begin{cases} kU_* y \, (1 - y/y_m), & \Delta \leq y \leq \alpha y_m \\ & \\ kU_* y_m \, (y/y_m - 1)^2 + kU_* y_m \, (2\alpha - 1)(1 - \alpha), & \alpha y_m < y < y_m \end{cases} \quad (8.1)$$

in which y is the vertical distance, counted from the bottom, y_m is the vertical distance of the point, at which the flow velocity is the maximum, Δ the thickness of laminar sublayer, α is a constant, which is taken as 0.8 based on experiment data.

The contribution of coarse particles moving as bed load on momentum exchange coefficient ε_m is obvious in the near bed region. It can be considered by adding a supplementary term in the first part of Equation (8.1) (Wang, 1984) as follows:

$$K_1 D^2 \, \frac{du}{dy}$$

Here D is the diameter of sediment, K_1 a constant. It means that in the near bed region the momentum exchange is intensified by bed load movement.

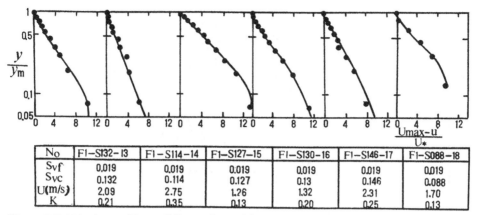

Figure 8.7. Velocity profiles at different fine particle concentration.

The existence of fine particles weakens the mentioned above effect. It can be considered by introducing a minus term:

$$-f(S_{vf})\,D^2\,\frac{du}{dy}$$

$f(S_{vf})$ is an unknown function. Later Song found $f(S_{vf}) = 10^4\,S_{vf}$ and $K_1 = 10$ based on experiment data.

So for flow carrying both coarse and fine particles one obtains:

$$\varepsilon_m = \begin{cases} kU_*y(1 - y/y_m) + (K_1 - f(S_{vf})D^2\,\dfrac{du}{dy}, & \Delta \leq y \leq \alpha y_m \\[2mm] kU_*y_m(1 - \alpha)(2\alpha - 1) + kU_*y_m(y/y_m - 1)^2, & \alpha y_m < y < y_m \end{cases}$$

(8.2)

And the velocity profile can be deduced from Equation (8.2) and the equation of shear stress. The result is:

$$\frac{U_{max} - U}{U_*} = \begin{cases} \dfrac{1}{K}\ln\dfrac{1.6}{\sqrt{3}}\dfrac{y_m}{y} + \dfrac{5(1 - 1000\,S_{vf}^3)}{K^3/\gamma_m}\left(\dfrac{5}{4} - \dfrac{y_m}{y}\right)\left(\dfrac{9}{4} + \dfrac{y_m}{y}\right), \\[2mm] \hspace{4cm} \Delta \leq y \leq \alpha y_m \\[2mm] \dfrac{1}{K}\ln\left(1 + \dfrac{25}{3}\left(1 - \dfrac{y}{y_m}\right)^2\right)^{1/2}, \hspace{1cm} a y_m \leq y \leq y_m \end{cases}$$

(8.3)

In Figure 8.7 the measured velocity profiles (dots) are compared with Equation (8.3) (curves). The agreement is satisfied.

8.5 SEDIMENT CARRYING CAPACITY

Wan & Sheng (1974) found that, in the region of high concentration the correla-

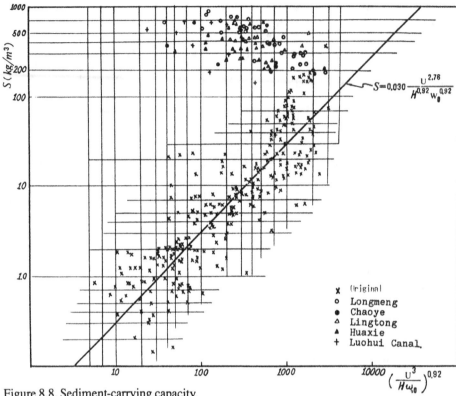

Figure 8.8. Sediment-carrying capacity.

tionship $S \sim U^3/gH\omega_o$, which is used in describing sediment carrying capacity, has a reverse tendency and it has a hook-like outline, as shown in Figure 8.8. Here S is the average concentration under equilibrium condition, ω_o is the fall velocity of a single particle in standstill clear water.

Cao (1975) treated the same problem in another way. Instead of ω_o, ω which is the gross fall velocity of particles at concentration S_v (in volume), was used in a similar plot $S_v \sim U^3/gH\omega$, see Figure 8.9. In that figure the reverse tendency disappeared.

Combining these two figures, we can deduce the following concept:

1. As shown in Figure 8.8, in the region of high concentration (about $S > 200$ kg/m^3) more sediment can be carried by flow with even weaker intensity. In other words, high concentration does not require high flow intensity to carry it. This is a very useful concept in practise.

2. The reason causing the reverse tendency of $S \sim U^3/gH\omega_0$ correlationship is the obvious reduction of the fall velocity at high concentration, particularly, at high concentration of fine particles. If the reduction of fall velocity due to concentration has been taken into consideration, the hyperconcentrated flow follows the same law as that followed by an ordinary sediment-laden flow.

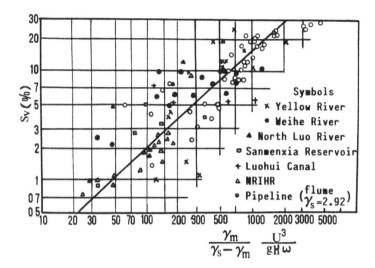

Figure 8.9. Revised sediment-carrying capacity.

$$\frac{\gamma_m}{\gamma_s - \gamma_m} \quad \frac{U^3}{gH\,\omega}$$

Table 8.1. Properties of plastic beads.

Density ρ_s	Volumetric diameter D	Fall velocity ω_0	Shape
1.34 g/cm^3	2.29 mm	12.3 cm/s	Cylindrical

Qi (1981) also discussed the effect of fine particles on coarse sediment carrying capacity. Four kinds of coarse sediment were put in clear water or in mud with different clay concentration and fall velocities of coarse particles were measured. Fall velocities decreased with the increase of clay concentration. Based on this phenomenon, he explained why the sediment carrying capacity increased with the increase of clay concentration. But no quantitative relationship was given.

Systematic study was carried out in flume experiments by Wan & Song (1987), which have been mentioned above. The tilting flume is 32 m long, 0.45 m wide and 0.60 m high. Plastic beads, properties of which are listed in Table 8.1, were used as coarse particles, Huayuankou clay was used as fine particles.

The bottom of the flume was covered by plastic beads and let clear water and then mud at different concentrations run over it. At same flow intensity the transport rate of coarse particles increased by dozen to nearly hundred times with the increase of clay concentration S_{vf}, see Figure 8.10. There R_b the hydraulic radius with respect to the bed, ω_0 is the fall velocity of a single coarse particle in clear water. In the figure, instead of H, R_b was used to consider the effect of walls.

In the figure there are two kinds of dots. For $S_{vf} < 0.0755$ dots are distributed along straight lines with slope 4/3. At same $U^3/g\,H\omega_0$ dots with higher S_{vf} are located higher. It means that at same $U^3/gH\,\omega_0$ more coarse sediment can be transported if S_{vf} is higher. For $S_{vf} > 0.0755$ dots are distributed along straight

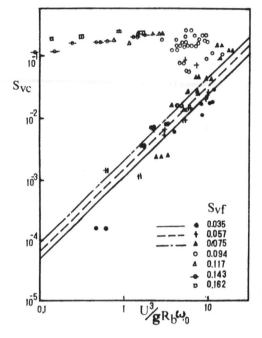

Figure 8.10. $S_{vc} - U^3/gR_b\omega_0$.

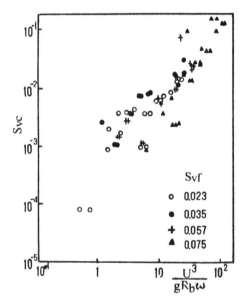

Figure 8.11. $S_{vc} - U^3/gR_b\omega$.

lines with much gentler slope. It means much more coarse particles can be transported and S_{vc} little varies with $U^3/gH\omega_0$. $S_{vf} = 0.0755$ is the concentration, at which coarse particles (plastic beads) do not settle according to Equation (5.21). In fact it could be observed that even at clay concentration $S_{vf} = 0.162$ coarse particles still settled under low intensity condition. The settlement might be caused by the fact that the flocculent structures in running slurry was partly destroyed and Bingham stress consequently reduced.

Considering the analysis in Section 8.3, one can image that the increase of sediment-carrying capacity with the increase of S_{vf} is the result of the reduction of settling velocity due to concentration. For $S_{vf} < 0.0755$ the group fall velocity ω of coarse particles in Bingham fluid with concentration S_{vf}, which is ascertained by the $C_D - \text{Re}_2$ curve in Figure 5.10, is used to replace ω_0 and Figure 8.10 is replotted as Figure 8.11. In Figure 8.11 dots of different S_{vf} mix together and form an unique tendency.

The problem of sediment-carrying capacity can also be treated from the view of energy consumption. The energy for suspending sediment particles directly originates from turbulence energy, which transforms from potential energy. Therefore there must be some kind of relationship between the suspension energy and the potential energy. As an average, the potential energy of turbid water with unit volume in unit time is $\gamma_m U J_e$, and the energy consumed in unit time for suspending particles in slurry with unit volume is $(\gamma_s - \gamma_m) S_{vc} \omega$. If the efficiency of energy transformation is e, the following formula can be written down:

$$(\gamma_s - \gamma_m) S_{vc} \omega = e \gamma_m U J_e$$

$$S_{vc} = e \frac{\gamma_m U J_e}{(\gamma_s - \gamma_m) \omega} \tag{8.4}$$

where ω is the group fall velocity of coarse particle in slurry.

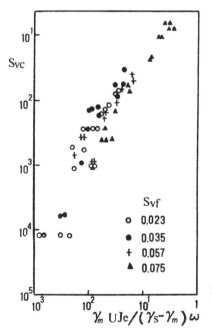

Figure 8.12. $S_{vc} - \gamma_m U J_e / (\gamma_s - \gamma_m) \omega$.

For experiments of $S_{vf} < 0.0755$ the

$$S_{vf} \sim \frac{\gamma_m U J_e}{(\gamma_s - \gamma_m)\omega}$$

correlationship is plotted in Figure 8.12. Dots of different concentration mix together and form an unique trend too.

For experiments of $S_{vf} > 0.0755$ coarse particles can keep stationary in slurry and ω should be zero according to Equation (2.13). In such case neither $S_{vc} \sim U^3/gR_b\omega$ nor $S_{vc} \sim \gamma_m U J_e/(\gamma_s - \gamma_m)\omega$ correlationship can be plotted. Although in fact the settlement of coarse particles can still be observed. As ω is so small in this case that the value of the Rouse number

$$Z = \frac{\omega}{kU_*}$$

is smaller than 0.06, the coarse particles become wash load (see Chapter 2.7). S_{vc} mainly depends on incoming concentration and the concept of sediment carrying capacity is not valid in this case.

8.6 FLOW RESISTANCE

A set of experiments with mixture of water and pure sand was carried out by Wang & Qian (1984). As a succession, experiments with mixture of water, sand and clay of three different concentrations were carried out by Song et al. (1986). These experiments have been described in Section 8.3. The main parameters in these experiments are listed in Table 8.2.

Experiment results of flow resistance are shown in Figure 8.13, in which the ordinate J_m is the pressure gradient counted in column of slurry, U is the average velocity. In the figure the straight line 1 is the resistance curve for clear water. The dotted line, which has a hook-like outline, is the resistance curve for the mixture of water and pure sand. It can be seen from the figure that in the range of $U > 2.5$ m/s the flow resistance increases with S_{vf} increasing. In the range of low velocity the flow resistance for the mixture of water and pure sand is the highest, it

Table 8.2. Main parameters in experiments.

Authors	Coarse particles				Fine particles				Velocity
	Sample	D_{50} (mm)	D_{max} (mm)	S_{vc}	Sample	D_{50} (mm)	D_{max} (mm)	S_{vf}	U (m/s)
Wang & Qian	Sand	.15	.40	.019 ~ .38				0	.50 ~ 5.0
Song et al.	Sand	.15	.40	.015 ~ .330	Clay	.0045	.074	.018	.50 ~ 3.7
								.030	
								.060	

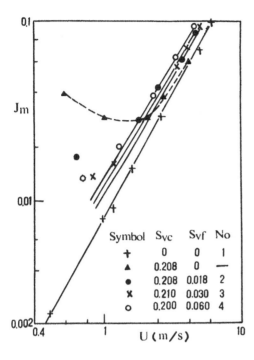

Symbol	S_{vc}	S_{vf}	No
+	0	0	1
▲	0.208	0	—
●	0.208	0.018	2
✕	0.210	0.030	3
○	0.200	0.060	4

Figure 8.13. $J_m - U$ correlationship.

decreases with S_{vf} increasing. The resistance for $S_{vf} = 0.03$ is the lowest. The resistance turns to increase as S_{vf} further increases.

Generally speaking, resistance of a sediment-laden flow consists of three parts: viscous resistance, turbulent resistance, and resistance caused by bed load movement and bed configuration (ripples and dunes). For pseudo-one-phase flow (pure clay slurry) only the first two parts exist. Under such conditions the relationship between friction factor f and Reynolds number Re_1 is shown in Figure 8.14 and discussed in Chapter 6. For flow with both fine and coarse particles the situation is different in different cases. The energy consumed on suspending sediment particles originates from turbulence energy, which finally dissipates and transforms into heat even in clear water. Therefore, no additional energy is needed for maintaining suspended load. As a result, the relationship between f and Re_1 for a slurry carrying suspended sand is the same as that for a slurry without suspended sand. Bed load is characterized by a relative motion between the particles and the surrounding fluid in both vertical and horizontal directions. Mutual collision between particles and collision between particles and bed occur intermittently. Consequently, certain energy is substracted from the potential energy of the flow for maintaining the movement of bed load particles. In case of existence of bed configuration vortex behind ripples and dunes consumes energy too. This means that additional energy is needed for bed load movement.

Results of experiments above mentioned are plotted in Figure 8.14, in which experiments with different S_{vf} are plotted with different symbols. Based on the

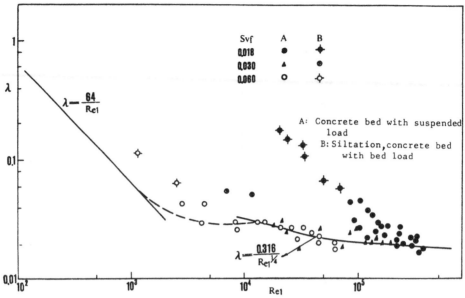

Figure 8.14. $f - Re_1$ correlationship for flow carrying both fine and coarse particles.

tendency of the experimental dots, the following information is obtained:

1. For two sets of experiments with $S_{vf} = 0.03$ or 0.06, the experimental dots are distributed along the curve for pseudo-one-phase flow, provided all sediment particles were suspended, even if in some experiments concentration of coarse particles was unevenly distributed. The concentration gradient was particularly obvious in experiments with $S_{vf} = 0.03$.

2. Whenever there was visible bed load movement or deposition on the bed, experimental dots are located significantly higher than the curves for pseudo-one-phase flow.

3. For experiments with $S_{vf} = 0.018$, besides the dots with deposition or bed load movement, some dots with median velocity are also located above the curve of pseudo-one-phase flow. It can be explained as follows: although judged by naked eyes, all the coarse particles were suspended and there was no bed load movement near the bed. Actually, the fall velocity of coarse particles was still large as S_{vf} was low, and some coarse particles near the bed intermittently collided with the bottom and jumped up. Jumping up, they were accelerated by the flowing water in the flow direction. Both collision and acceleration cost additional energy. They can be considered to be in a transitional state between suspension and bed load movement.

Thomas (1978) also put forward the following idea: most industrial slurries possess both coarse and fine particles and so can be envisaged as coarse particles suspended not in water, but in a fine particle slurry or heavy medium. Replacing

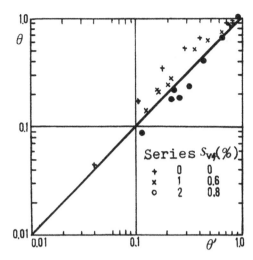

Figure 8.15. $\theta - \Theta'$.

water with a heavy medium reduces the settling tendency of heavy particles and enhances the transport of heavy particles.

Dai & Qian (1982) clearly illustrated the dual effects played by fine particles. On the one hand, the fall velocity of coarse particles reduces due to the existence of fine particles and more coarse particles moving originally as bed load transform into suspended load. Consequently, flow resistance reduces. On the other hand, the viscosity of fluid and the flow resistance consequently increase due to the existence of fine particles. It is particularly obvious in laminar flow region and transitional region.

Flow resistance is also indirectly influenced by the adding of fine particles through the change of bed configuration. As mentioned in the Section 8.2, compared with those in clear water, dunes in clay slurries have gentler and smoother outline. The higher the clay concentration, the larger the difference. Therefore, in clay slurries form resistance reduces, and consequently total resistance reduces too.

In Wan's experiments, which was mentioned in the Section 8.2, dimensionless parameters θ, θ'' and θ' are used to represent the total flow resistance, resistance due to bed form and resistance due to skin friction, respectively. The definition of these parameters is as follows:

$$\theta = \frac{R_b J}{(s-1)D} \tag{8.5}$$

$$\theta' = \frac{R_b' J}{(s-1)D} \tag{8.6}$$

$$\theta'' = \frac{R_b'' J}{(s-1)D} \tag{8.7}$$

in which J is the slope, $s = \rho_s/\rho$ specific density of particles, R_b hydraulic radius with respect to bed, R_b' hydraulic radius with respect to the skin friction of the bed, R_b'' hydraulic radius with respect to form resistance of the bed.

$$R_b = R_b' + R_b'' \tag{8.8}$$

It means the total resistance is composed of skin friction and form resistance. Based on experiment data Figure 8.15 is plotted.

It can be seen from the figure that at the same θ, θ' decreases with increasing S_{vf}. It means that the total resistance reduces as clay concentration increases. The reduction of total resistance is caused by the disappearance or the change of the bed configuration.

8.7 DRAG REDUCTION

Having discussed flow resistance of pseudo-one-phase flow in Chapter 6 as well as that of flow carrying both fine and coarse particles in last section, we turn to discuss the problem of drag reduction. That is, is the flow resistance of a sediment-laden flow lower than that of a clear water flow, or not?

Before this discussion it is necessary to ascertain the condition of comparison. That is, under which condition the comparison is conducted. Someone compares the flow resistance of a sediment-laden flow with that of a clear water at same Reynolds number. Due to the higher viscosity of the sediment-laden flow, at same Reynolds number the sediment-laden flow always has a higher velocity than a clear water flow does. It means that such comparison is carried out between two flow with different velocity or discharge.

From the view of engineering practise, it is reasonable to compare the resistance under the condition of same flow velocity. That is, is the resistance of a sediment-laden flow lower than that of a clear water flow with the same velocity? or *vice verse*?

We will compare a pseudo-one-phase flow, that is, a flow with pure fine particles with a clear water flow at first, as we discussed in Chapter 6, provided a revised Reynolds number Re_1 is used, the flow resistance of a pseudo-one-phase flow can be described by the $f - Re_1$ relationship of the clear water flow except in the transitional region. In this region f of a pseudo-one-phase flow is a little lower than that of a clear water flow.

As the viscosity of a pseudo-one-phase flow is always higher than that of a clear water flow, the revised Reynolds number Re_1 of a pseudo-one-phase flow is always smaller than the Reynolds number Re of a clear water flow with same velocity. Both in laminar flow region and in turbulent flow region with hydraulically smooth boundary, the $f - Re_1$ relationship has a minus slope, see Figure 6.13. So a smaller Re_1 means a larger f. In other words, in these regions the resistance of a pseudo-one-phase flow (sediment-laden flow) is higher than that of a clear water flow. In fully developed turbulent flow region f keeps constant and has nothing to do with Reynolds number Re_1 and the resistance of a pseudo-one-phase flow is the same as that of a clear water flow. Only in the transitional flow region the resistance of a pseudo-one-phase flow might be lower than that of a clear water flow.

As a summary, in most cases a pseudo-one-phase flow (sediment-laden flow) has a higher resistance than a clear water flow does. Only in the transitional flow region a pseudo-one-phase flow might have a lower resistance.

The situation is different for flow carrying both coarse and fine particles. Provided coarse particles are already fully suspended in clear water flow, adding fine particles in the flow does not cause any reduction of resistance. On the contrary, provided coarse particles partly or all move as bed load or deposit, adding fine particles makes the fall velocity of coarse particles reduce. Consequently, particles originally moving as bed load turn to be suspended load, and bed configuration becomes smoother and gentler. All these cause drag reduction. As a result, drag reduction is associated with the adding of fine particles.

As shown in figure 8.14, at low clay concentration ($S_{vf} = 0.018$), bed load occurred and the friction factor f was larger. At high clay concentration, for instance, $S_{vf} = 0.060$, bed load turned to be suspended load and the friction factor f dropped consequently.

REFERENCES

Cao, R. 1975. Discussion on sediment-carrying capacity of hyperconcentrated flow (in Chinese). *Selected Papers of The Symposium on Sediment Problems on The Yellow River*, Vol. 2: 249-258.

Dai, J. & N. Qian (Ning Chien) 1982. The effect of sediment composition and content of fine particles on the characteristics of two-phase flow in pipes (in Chinese). *J. of Sediment Research*, No. 1: 14-23.

Hisamitsu, N., Y. Shoji & S. Kosugi 1978. Effect of added fine particles on flow properties of settling slurries. *Proc., Hydrotransport* 5, pp. D3.

Kazanskij, I. & H. Bruhl 1972. Influence of high concentrated rigid particles on macroturbulence characteristics of pipe flow. *Proc., Hydrotransport* 5, pp. D7 91-102.

Kikkawa, H. & S. Fukuoka 1869. The characteristics of flow with wash load. *Proc. 13th Cong. Intern. Assoc. Hyd. Res.*, Vol. 2: 233-240.

Qi, P. 1981. Preliminary study on the effect of finest sediment content on the sediment carrying capacity of the Yellow River (in Chinese). *J. of Sediment Research*, 3:81-88.

Qian, N. (Ning Chien), R. Zhang, X. Wang & Z. Wan 1984. The hyperconcentrated flow in the main stem and tributaries of the Yellow River. *Proc. of International Workshop on Flow at Hyperconcentrations of Sediment*. Publication of IRTCES.

Sakamoto, M., M. Mase, Y. Nagawa, K. Uhida & Y. Kamino 1978. A hydraulic transport study of coarse materials including fine particles with hydrohoist (in Chinese). *Proc. Hydrotransport* 5, pp. D6.

Song, T. 1986. Experimental study on velocity profiles of hyperconcentrated flow in pipes (in Chinese). *J. of Sediment Research* 3:53-61.

Song, T, Z. Wan & N. Qian (Ning Chien) 1986. The effect of fine particles of the two-phase flow with hyperconcentration of coarse grains (in Chinese). *J. of Hydraulic Engineering* 4:1-10.

Thomas, A.D. 1978. Coarse particles of heavy medium – turbulent pressure drop reduction under laminar flow. *Proc. Hydrotransport* 5, pp. D5.

Wan, Z. 1982. Bed material movement in hyperconcentrated flow. Series Paper No. 31, Institute of Hydrodynamics and Hydraulic Engineering, Technical University of Denmark.

Wang, Z. 1984. Experimental study on mechanism of hyperconcentrated flow. Doctoral thesis (in Chinese). Institute of Water Conservancy and Hydroelectric Power Research.

Wang, Z. & N. Qian (Ning Chien) 1984. Experiment study of two-phase turbulent flow with hyperconcentration of coarse particles. Scienta Sinica (Series A), 27(12):1317-1327.

Wan, Z. & S. Shen 1974. Hyperconcentrated flow on the Yellow River and its tributaries (in Chinese). *Selected papers of the Symposium on Sediment Problems on the Yellow River*, 1:141-158.

Wan, Z. & T. Song 1978. The effect of fine particles on the vertical concentration distribution rate of coarse particles (in Chinese). *J. of Hydraulic Engineering*, 8:20-31.

CHAPTER 9

Characteristics of fluvial processes associated with hyperconcentrated flow

Fluvial processes associated with hyperconcentrated flow are usually character-ized by great and rapid change of aggradation or degradation. The volumetric change of a hyperconcentrated flood due to gigantic aggradation or degradaion might be massive and therefore volumetric change caused by aggradatiion or degradatiion should be taken into account. The morphological change of the river pattern during the passage of a hyperconcentrated flood can also be significant, and it affects the sediment transport in turn. Peculiar forms of siltation or erosion associated with hyperconcentrated flow are clogging or ripping up the bottom. Both of them will be described and analysed in this chapter.

9.1 LARGE AMPLITUDE AND HIGH SPEED OF AGGRADATION OR DEGRADATION

Rapid siltation is quite often associated with hyperconcentrated floods in some tributaries or on the flood plain of the Yellow River, as mentioned above. In Figure 9.1 is an example of such siltation recorded at a gauging station on Luhe River, which is a second order tributary of the Yellow River (Wan, 1982). The hydro-graphs of discharge, concentration and water level at the gauging station is shown on the right diagram (9.1b). The variation of that cross section is shown in Figure 9.1a. The ordinal number and the time of sounding are also shown on both diagrams. The river bed, as well as water level, rose rapidly during the recession of the flood.

The hyperconcentrated flood is not always associated with rapid siltation. On the contrary, in some cases vigorous erosion happens while the hyperconcentrated flood passing, as shown in Table 1.2 (Wan & Sheng, 1978). The elevation of the average river bed, that of the thalweg bed, as well as the water stage lowered in phase, as shown in Figure 9.2 (Qian et al., 1979). Similar phenomena have happened on the Weihe River, North Luohe River, etc. Another example happen-ing on North Luohe River is plotted in Figure 9.3.

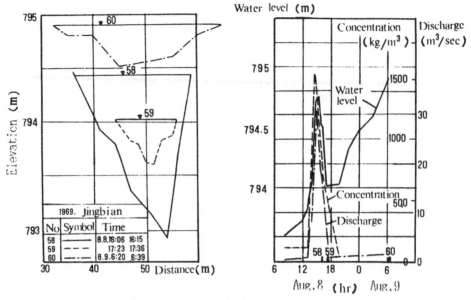

Figure 9.1. The 'clogging' phenomenon on the Luhe River.

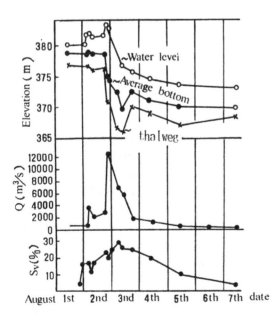

Figure 9.2. The serious erosion at Longmen Gauging Station, on the Yellow River.

Such degradation takes its peculiar form of erosion and is called as 'ripping up the bottom' by local habitants. More detailed description of such phenomenon and also an explanation of its mechanism will be given in Section 9.5.

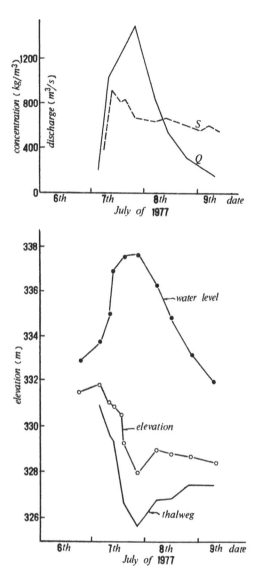

Figure 9.3. The serious erosion at Chaoyi Gauging Station, on the North Luohe River.

9.2 VOLUMETRIC CHANGE OF A HYPERCONCENTRATED FLOOD CAUSED BY RAPID AND SERIOUS DEGRADATION OR AGGRADATION

Rapid siltation or vigorous erosion causes some anomalies in stage and discharge. The most striking example happened in the lower reaches of the Yellow River in August, 1977. In Figure 9.4 hydrographs of stages at several gauging stations located between Xiaolangdi Gauging Station and Huayuankuo Gauging Station are plotted. The distances between these stations and Xiaolangdi Gauging Station

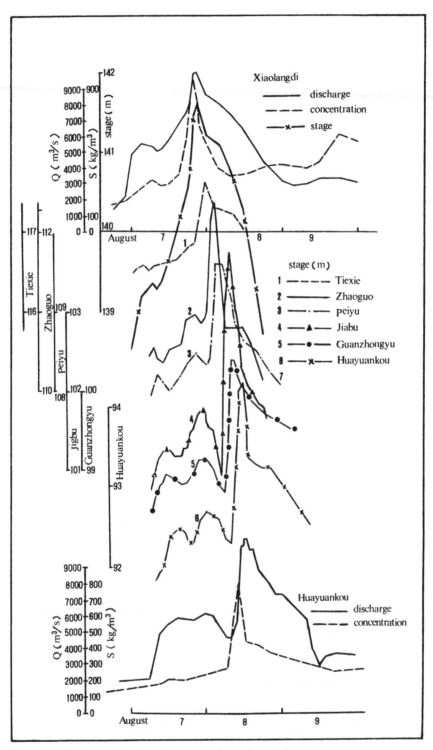

Figure 9.4. Hydrograph of stages at several gauging stations.

Table 9.1 Distances from the Xiaolangdi Gauging Station.

Station	Tiexie	Zhaoguo	Peiyu	Jiabu	Guanzhuangyu	Huayuankou
Distance (km)	26	48	59	91	100	135

are listed in Table 9.1. (Qi & Zhao, 1985). No flood came from tributaries in this reach in this period.

Corresponding to the rising period of the flood at Xiaolangdi with rapid ascent of sediment concentration, the water level downstream from Zhaoguo did not rise, but dropped down at first. Then it was followed by a rapid rising. The water level rose for 2.48 m and 2.04 m within two hours at Jiabu and Huayuankou respectively. Besides, the peak discharge at Xiaolangdi was 10 100 m^3/s. Without incoming flow from tributaries, the peak discharge at Huayuankuo did not decrease as usually happens due to flood transformation, but increased to 10 800 m^3/s instead. Such variation of stage and discharge seems anomalous and unexpected.

Qi & Zhao (1985) analysed the anomaly and gave an explanation. Different from that of an ordinary sediment-laden flow, the volumetric change of a hyperconcentrated flow caused by rapid and serious degradation or aggradation can not be ignored. Generally speaking, the unit weight of river deposit is about 1.4 t/m^3. Provided the concentration of a hyperconcentrated flow is in the range of 400-1000 kg/m^3 and all the sediment deposites, the discharge at a downstream gauging station will diminish by 29-72% because of the reduction of the volume of settled sediment together with the water in pores between sediment particles. The higher the concentration, the more obvious the variation.

Taking the propagation time for a flood from Xiaolangdi to Huayuankou as 13 hours and shifting the hydrograph at Xiaolangdi for 13 hours, they overlapped the hydrograph of sediment flux at Xiaolangdi and that at Huayuankou for the flood in August 1977, see Figure 9.5. It reveals that severe aggradation took place in this river reach during the period of flood rising. Take the unit weight of deposition as 1.4 t/m^3. The increment of discharge ΔQ caused by degradation or aggradation should be $\Delta Q_s/1.4$, in which ΔQ_s is the variation of sediment transport rate. Superimpose the increment of discharge ΔQ on the original observed flood hydrograph at Huayuankou for corresponding time interval, they got the revised flood hydrograph of discharge as shown in Figure 9.5. This revised hydrograph of discharge is essentially consistent with the hydrograph of discharge at Xiaolangdi, and the apparent reduction of river discharge during rising period no longer exists. This fact proves that the so called anomaly is caused by volumetric change due to serious aggradation or degradation, which is usually neglected in a common sediment-laden flow. The reason for abrupt rise in river stage was caused by rapid increase of discharge. With the same amplitude of increase in discharge, the rise of stage did not differ appreciably from that in other cases.

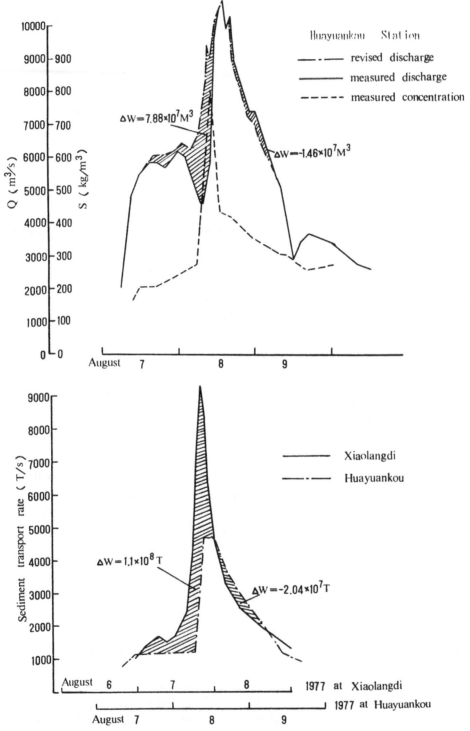

Figure 9.5. Comparison of hydrographes at Huayuankou and Xiaolangdi.

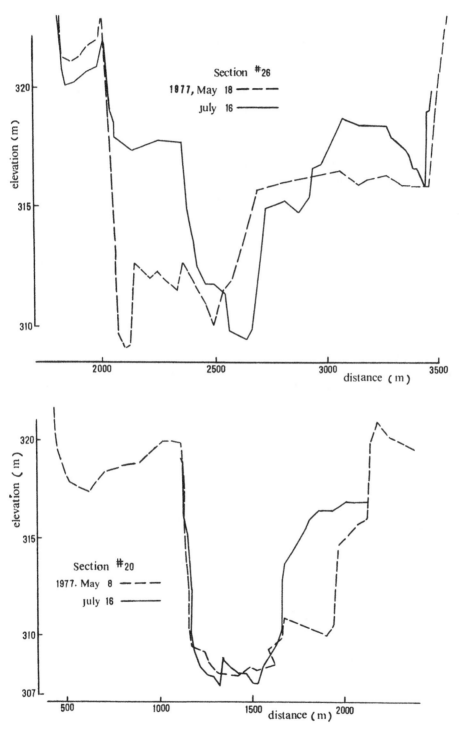

Figure 9.6. Change of cross sections in Sanmenxia Reservoir caused by the passage of a hyperconcentrated flood on July 7, 1977.

Besides, at some places such as Jiabu flow ran against the bank with a large attack angle. And it caused very high local water stage.

9.3 CHANGE OF CROSS SECTION AND MODIFICATION OF CHANNEL PATTERN CAUSED BY HYPERCONCENTRATED FLOW

The striking feature of the change caused by the passage of a hyperconcentrated flow is the narrowing and deepening of the cross section. It is caused by the accretion of the massive lateral berms to the sides of the channel and the simultaneous degradation of the channel bottom. Such feature is displayed not only in rivers, but also in canals and in reservoirs. Figure 9.6 shows two cross section in Sanmenxia Reservoir. The change caused by the hyperconcentrated flood on July 7 is distinct (Qi & Zhao, 1987).

Similar change happens quite often on the Lower Reaches of the Yellow River. In Figure 9.7 is the comparison of the pre- and post-flood cross section at Huayuankou. The hydrograph and the variation of the elevation of the average channel bed and thalweg are shown in Figure 9.8 (Qi & Zhao, 1985).

Janda & Meyer (1985) studied the channel morphology changes caused by debris flows and hyperconcentrated streamflows and drew the same conclusion. Debris flows and hyperconcentrated streamflows in the North Fork and main stem of the Toutle River leave a channel narrower, straighter, and with finer bed

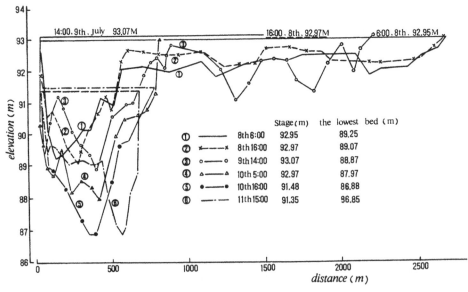

Figure 9.7. Change of cross section at Huayuankou caused by the passage of a hyperconcentrated flood in July, 1977.

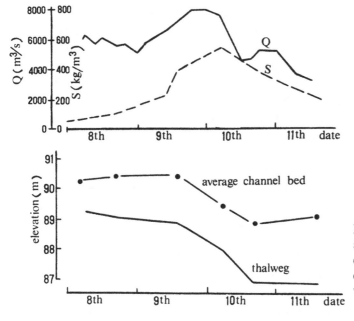

Figure 9.8. Hydrograph and the variation of the elevation of the average channel bed and thalweg point.

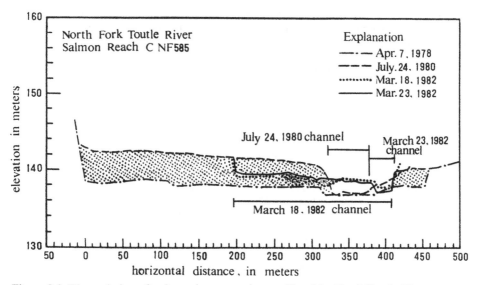

Figure 9.9. The variation of a channel cross section profile of the North Toutle River.

material than sediment-laden flows of comparable discharge do. As an example, a channel cross section profile of the North Fork Toutle River is shown in Figure 9.9. The light shading represents North Fork Toutle lahar deposits on May 18, 1980. The darker shading represents hyperconcentrated flow deposits on March 19, 1982.

Not only floodplains but also by-channels are silted up while a hyperconcen-trated flood is passing through. It is particularly distinct on the lower reaches of the Yellow River in Henan Province. In this reach Yellow River is usually a wandering stream whereas unstable central islands are dispersed and channels are in the process of everlasting wandering. During the passage of a hyperconcen-trated flood floodplains and by-channels are silted and blocked while the main channel is deepened. An original wandering stream with dispersed central islands and channels turns into a meandering stream with a single curved, deep and narrow channel.

The North Luohe River is a second order tributary originating from the northeast of the loess plateau. Floods on this river are always at hyperconcentra-tions with very few exceptions. Discharge of base flow is usually very little. So the river is mainly moulded by hyperconcentrated flood and has a meandering narrow, deep channel with a width-depth ratio $\sqrt{B}/H$ of about 14.

The Weihe River is a first order tributary of the Yellow River. Tributaries from its north part, including Jinghe River, are heavily sediment-laden ones, and those from south part are relatively clear ones. Whenever floods from Jinghe and other northern tributaries, which usually occur in July and August, propagate the lower reaches of the Weihe River, the latter becomes narrow and deep. Then in September or October when floods from south tributaries come the river is widened again. Morphological change of the river channel influences its hydrau-lic characteristics as well as the sediment transport capacity of the flow along it. It is found that the flood propagation speed is raised and the flood propagation time

Table 9.2. The variation of propagation time of hyperconcentrated floods.

Year	No. of flood	Reach	Time	Peak discharge Q_{max} (m³/s)	R^*	S (kg/m³)	Propagation time (hr)	speed (m/s)	S_{max} (kg/m³)
1973	1	Xiaolangdi	8.27. 1:42	4230	1.11	110	33.3	1.07	110
		Huayuankou	8.28.11:00	4710		120			150
	2	Xiaolangdi	8.30. 0:00	3630	1.38	360	22.0	1.60	500
		Huayuankou	8.30.22:00	5020		230			150
	3	Xiaolangdi	9.2. 12:00	4400	1.34	325	22.0	1.60	338
		Huayuankou	9.3. 10:00	5890		330			348
1977	1	Longmen	7.6. 17:00	14500	0.93	575	13.0	2.80	690
		Tongguan	7.7. 6:00	13600		615			616
	2	Longmen	8.3. 5:00	13600	0.88	145	10.0	3.61	551
		Tongguan	8.3. 15:00	12000		185			235
1977	3	Longmen	8.6. 15:30	12700	1.21	480	7.50	4.81	821
		Tongguan	8.6. 23:00	15400		911			911
1977	1	Xiaolangdi	7.8. 15:30	8100	1.00	170	28.5	1.25	535
		Huayuankou	7.9. 19:00	8100		450			546
	2	Xiaolangdi	8.7. 21:00	10100	1.07	840	15.7	2.26	941
		Huayuankou	8.8. 12:42	10800		437			809

*R is the ratio of the peak discharge at Xiaolangdi to that at Huayuankou.

Table 9.3. The change of sediment-carrying capacity in a hyperconcentrated flood in 1973.

Time	Hydraulic and sediment factor at Sanmenxia			In the reach between Sanmenxia and Jiahetan	
	Q_{max} (m³/s)	S_{max} (kg/m³)	D_{50} (mm)	Deposition (10⁸t)	Releasing ratio (%)
August 28-30	3630	509	.04-.05	1.32	66
Sept. 1-3	4400	338	.04-.05	−0.594	124

is shortened as a wandering river turns to be a meandering river with a single, narrow and deep channel while a hyperconcentrated flood is passing through. In Table 9.2 propagation speed and propagation time of different hyperconcentrated floods passing through Longmen-Tongguan reach or Xiaolangdi-Huayuankou reach are compared (Qi et al., 1984). In general case the Yellow River in both reaches is wandering one. In the same year the later hyperconcentrated flood propagates faster than the earlier ones. It is just because a smooth, meandering single channel shows lower hydraulic resistance and flow runs faster.

The effect of morphological change on sediment transport capacity is also obvious. The hyperconcentrated flood in 1973 lasted seven days. The morphological change associated with hyperconcentrated flood was fulfilled in the first several days. As shown in Table 9.3, serious siltation happened in the first four days and the corresponding releasing ratio in that period is only 66%. Here releasing ratio is the ratio of the outcoming sediment to the incoming sediment of that reach. In the next three days instead of siltation, erosion was associated with the hyperconcentrated flood, and the releasing ratio was raised up to 124%.

Narrow, deep channel favors the transport of hyperconcentrated flood. Based on field data from the stem and tributaries of the Yellow River, flow at concentration of 800 kg/m³ or more was conveyed over a long distance without siltation along some rivers with narrow and deep channel, such as Weihe River, North Luohe River, etc., under the condition of unit discharge larger than 5 m³/s-m and slope of 10^{-4} (Qi & Zhao, 1985).

9.4 CLOGGING

'Clogging' phenomenon has been recorded in some small tributaries of the Yellow River during the recession of a hyperconcentrated flood when the discharge is small but the concentration remains high. The whole river stops flowing for several minutes and meanwhile the stage rises, than it flows again and water level drops. Afterwards it stops again. Such intermittent phenomenon repeats for several times. Figure 9.10 shows an example taken at Lanxipu Gauging Station on Heihe River in July, 1967. The period is 3-7 minutes and amplitude of water level is 15-26 cm. The flow parameters obtained by twice measurements as marked in the figure are listed in Table 9.4. In the table Fr is the Froud number.

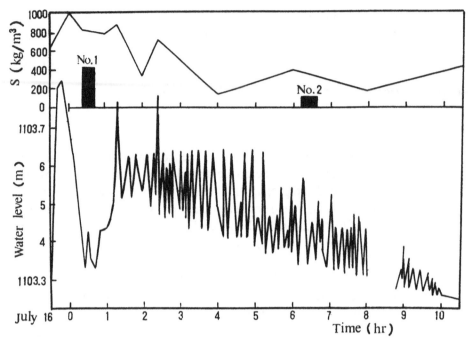

Figure 9.10. Unstable phenomenon of hyperconcentrated flow at Lanxipu Gauging Station, Heihe River.

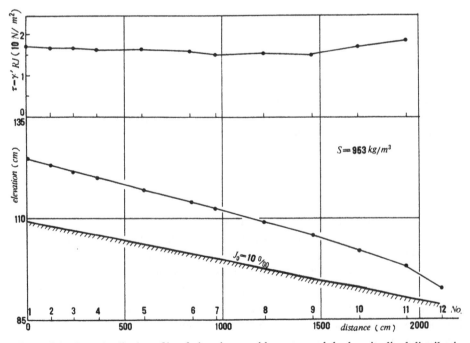

Figure 9.11. Longitudinal profile of clogging muddy water and the longitudinal distribution of the shear stress acting on boundary.

Such phenomenon has been reproduced in laboratories (Wan et al., 1979; Engelund & Wan, 1984). Experiments were conducted in recirculating tilting flumes with mud or bentonite suspension, concentration, slope and discharge changed in different runs. 'Clogging' happens at different slopes only if the concentration of the flow is high enough. Whenever the valve at the inlet is shut off, clogging muddy water is formed. The longitudinal profile of the clogging muddy water has a convex outline, see Figure 9.11 (Wan et al., 1978). Under the condition of same bed slope, the higher the concentration, the higher the water

Table 9.4. Flow and sediment parameters in the period of unstable flow.

No	Discharge (m^3/s)	Average velocity (m/s)	Water depth (m)	Width (m)	Slope (10^{-3})	Fr
1	2.16	0.14	0.33	16		0.23
2	0.58	0.40	0.29	5	4.5	0.24

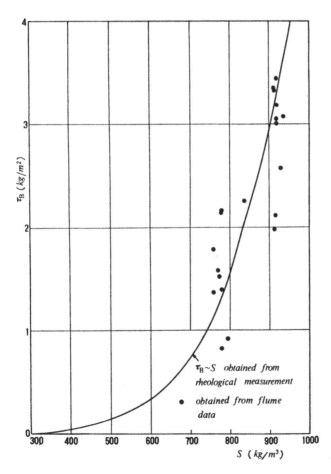

Figure 9.12. The comparison of the shear of a clogging muddy flow with the yield stress at corresponding concentrations.

surface. The longitudinal distribution of the shear $\gamma_m RJ$ acting on the boundary is shown in the upper part of the figure. Here γ_m is the unit weight of the muddy water, R hydraulic radius and J local water surface slope. It is clear that the shear stress keeps nearly constant along the distance. The shear stress in different experiments is compared with the yield stress corresponding to that concentration, which is got from rheological measurement, see Figure 9.12. They agree with each other as a tendency despite a certain scatter.

Based on the foregoing facts the following conclusion can be deduced: a hyperconcentrated flow clogs when the shear acting on the boundary equals its corresponding yield stress. Or in other words, a hyperconcentrated flow stops moving when the component of gravity along the slope is balanced by the resultant force of the shear stress acting on its boundary, which equals yield stress at that concentration.

During the decline of discharge, intermittent flow occurs. When the discharge decreases to a certain magnitude, the muddy water in the flume stops moving entirely at first. Because the little incoming flow which enters the flume from the inlet continues, the water level near the inlet of the flume rises gradually and the longitudinal slope of the water surface increases correspondingly. It can be deduced that the shear acting on the bounndary increases too. As the water level near the inlet rises to a certain degree, the muddy water in the flume flows for a while and the water level lowers. Then the muddy water stops flowing and the next circulation starts again. The variation of the stage at gauging stations No.7 (in the midst of the flume, see Figure 9.11), No.4 (near the inlet) and No.11 (near the outlet) is shown in Figure 9.13. Taking $\Delta H/H$ as a parameter reflecting the fluctuation intensity of the flow, we plotted $\Delta H/H$ versus the Reynolds number in

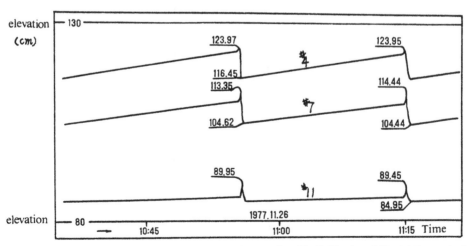

Figure 9.13. The variation of the stage at gauging station No.7, No.4 and No.11 (record of automatic level meter).

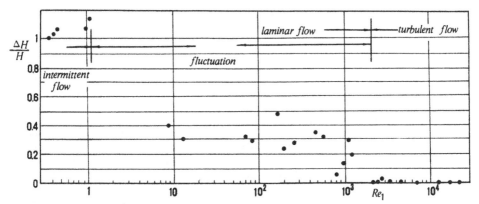

Figure 9.14. Different flow patterns in different Re_1 region.

Figure 9.14, in which H and ΔH are the water depth and the variation of the water depth at gauging station No.7. (Wan et al., 1979). It shows that fluctuation occurs under the condition of laminar flow ($Re_1 < 2000$). The fluctuation amplitude increases as the Reynolds number decreases. And intermittent flow occurs under the condition of $Re_1 < 1.1$. The definition of Re_1 is shown in Chapter 6.

9.5 'RIPPING UP THE BOTTOM'

'Ripping up the bottom' is a peculiar form of vigorous erosion associated with hyperconcentrations and large discharges. Such fantastic phenomenon has been recorded at Longmen, Tongguan on the middle reaches of the Yellow River, Lingtong on the Weihe River and once on the lower reaches of the Yellow River upstream from Huayuankou. Record of such phenomenon at Longmen and Lingtong is listed in Table 9.5.

'Ripping up the bottom' does not happen when the concentration is high but the discharge is not large enough. In the past people thought that after an issue of 'ripping up the bottom' another issue of 'ripping up the bottom' will not happen in a short period. But in 1977 'ripping up the bottom' happened in July at first, then it happened again in August.

'ripping up the bottom' is different from an ordinary erosion caused by common floods. The particular phenomenon of 'ripping up the bottom' is described by local witnesses as follow: as the flood passed, a block of the river bed, about one meter thick, was torn off by the flow; it turned around and stood up, towering over the water surface for several seconds. Then it collapsed with a great noise. After several minutes another block of the river bed towered over the water surface and collapsed again' (Wan, 1982).

The mechanism of such unusual erosion has been studied by several research workers. Wan & Sheng (1974) consider that 'ripping up the bottom' phenomenon

Table 9.5. Recorded 'ripping up the bottom' phenomenon at Longmen and Lingtong.

Guaging station	Year	Period of vigorous erosion			Erosion thick-ness (m)	S (kg/m^3)	Q (m^3/s)
		Start	Stop	Relative to peak discharge			
Longmen	1951	8.14.10:00	8.17.14:00	Before peak	2.6	35-542	2200- 5530
	1954	9.2.22:30	9.4.0:30	At peak	2.3	130-650	11800-16400
	1964	7.6.19:00	7.1. 6:30	After peak	3.1	350-610	1000- 8050
	1966	7.18.18:00	7.19.20:00	After peak	7.4	580-935	1000- 5300
	1969	7.27.22:00	7.28. 8:00	After peak	1.8	660-750	1500- 5000
	1970	8.2.19:25	8.3.10:00	Before peak	8.8	540-825	2980-13800
	1977	7.6.13:55	7.7.8:50	At peak	4.8	576-694	6890-11500
	1977	8.6.2:52	8.6.14:28	Before peak	2.0	821-	7580-12700
Lington	1964	7.11.-14:00	7.18.19:00		0.7	230-602	1000- 3120
	1966	8.13.20:00	8.14		0.5	400-670	2120- 3980
	1966	7.27.11:00	7.28.15:00		0.9	400-690	1800- 6200

happens provided the following two conditions are fulfilled:

1. The flow intensity is high enough to lift pieces of bed. It can be described by the criterion:

$$\frac{\gamma_m HJ}{(\gamma' - \gamma_m)\Delta} \gg \theta_c \qquad (9.1)$$

in which γ_m unit weight of the hyperconcentrated flow. γ' unit weight of the saturated bed deposit, H water depth, J slope, Δ thickness of pieces of bed, taken as 1 m preliminarily.

2. The concentration is so high that eroded sediment can be easily carried away by the flow.

Based on field data of Longmen gauging station, they tried to find out the critical values and concentration. But it should be pointed out that in general case the unit weight of the saturated bed deposit is about 1870 kg/m^3, which corresponds to a dry unit weight of 1400 kg/m^3. But Wan and Sheng took $\gamma'_m = 2270$ kg/m^3, which is too large, as the unit weight of the saturated bed deposit.

Later, Wan (1982) further explained why the bed was eroded in pieces, instead of in individual particles. On the one hand, the threshold velocity of individual particles in hyperconcentrated flow increases with increasing concentration. On the other hand, the submerged weight of pieces of bed decreases obviously with concentration increasing. It favors the lift of pieces of bed. As concentration increases to a certain degree, pieces of bed might be lifted earlier than individual particles.

Research workers at Wuhan Institute of Hydraulic and Electric Engineering (1982) also considered pieces of bed moving units. Criteria suggested by them are as follows:

1. Pieces of bed can be eroded. If the area and the thickness of a piece of bed is A and t, then the dynamic buoyancy acting on it is $C_2(At)^{2/3} C_1 \gamma_m V^2 /2g$, in which

V average velocity of the flow, g gravity, C_2 a coefficient considering the shape factor, C_1 a coefficient considering the intensity of dynamic bouyancy. The resistance is the submerged weight of that piece of bed.

That piece of bed can be lifted up provided the following condition is fulfilled.

$$C_2(At)^{2/3} C_1 \gamma_m \frac{V^2}{2g} > (\gamma' - \gamma_m)At$$

or

$$\frac{C_1 C_2 \frac{V^2}{2g}}{(At)^{1/3}} > \frac{\gamma' - \gamma_m}{\gamma_m} \qquad (9.2)$$

2. Eroded sediment can be carried away by the flow. The authors think that this condition can be fulfilled provided the flow is a hyperconcentrated turbulent one. No concrete figures are given.

The authors did not give the idea of the sizes A and t, and no magnitude of coefficients C_1 and C_2 is given too. So the criterion could not be examined.

Wang & Gu (1982) also studied the 'ripping up the bottom' phenomenon. They think that 'ripping up the bottom' phenomenon happens provided the following three conditions can be fulfilled.

1. *The shear stress* τ. Acting on bed by the flow is much larger than the threshold shear stress τ_c.

$$\tau \gg \tau_c \qquad (9.3)$$

The threshold shear stress τ_c is calculated according to authors' formula:

$$\tau_c = \tau_B + K_*(\gamma_s - \gamma_0)D \qquad (9.4)$$

in which γ_s, γ_0 the unit weight of sediment and that of water, respectively, D the diameter of deposit, τ_B yield stress of the deposit and can be related to the concentration of the deposit S_b by authors' formula:

$$\tau_B = 0.0064 \, S_b + 0.098 \, S_b^2 \qquad (9.5)$$

K_* in formula (9.5) is a parameter related to the median diameter D_{50} of the deposit and the porosity of the deposit ε, which is taken as 0.4.

$$K_* = 0.062 \times 10^{8(0.245 - 0.222 \lg D - \varepsilon)}$$
$$= 0.062 \times 10^{8(0.245 - 0.222 \lg D - 0.4)} \qquad (9.6)$$

2. *All the incoming sediment is wash load*. According to authors' definition, it can be written as:

$$\omega < VJ \qquad (9.7)$$

in which V the average velocity of the flow, J slope, ω average fall velocity of the incoming sediment at that concentration. The effect of concentration on fall velocity is considered by authors' formula:

$$\omega = \omega_0 (1 - \beta_* S) \qquad (9.8)$$

in which ω_0 the average fall velocity of incoming sediment settling in clear water, S the average concentration of the flow, β_* a parameter:

$$\beta_* = (0.775 + 0.222 \lg D_{50}) \qquad (9.9)$$

3. *The bed is a solid one*. The authors checked the aforementioned three criterion by field data of some hyperconcentrated floods associated with 'ripping up the bottom' phenomenon. The agreement is quite fair. It should be pointed out that the three criterion are easy to be met. Not only flow and sediment parameters of floods associated with 'ripping up the bottom' phenomenon, but also flow and sediment parameters of floods without 'ripping up the bottom' phenomenon meet the criterion. Some example of the latter are shown in Table 9.6.

Based on statistics of field data Miao & Fang (1984) gave the combination of the following two conditions as the criteria of 'ripping up the bottom' phenomenon:
1. Concentration higher than $400 \, \text{kg/m}^3$ lasts more than 10 hours.
2. Discharge more than $6000 \, \text{m}^3/\text{s}$ lasts more than 5-6 hours.
No mechanism or further explanation was given.

Later, Wang (1986) discussed this problem and suggested the following criterion:

$$\frac{\omega_b}{\kappa \beta V_*} < 0.06 \qquad (9.10)$$

in which ω_b the fall velocity of bed material at corresponding concentration S_v, V_* shear velocity during the flood, κ Karman constant, taken as 0.4, β coefficient, taken as 1.25. The meaning of the criteria is that eroded bed material turns to be wash load and can be carried away. The fall velocity ω_b is calculated according to author's suggestion:

$$\omega_b = \omega_{b0} (1 - S_v)^5 \qquad (9.11)$$

in which ω_{b0} is the fall velocity of bed material in clear water. The author also checked his criteria by field data. In his paper Wang mentioned that there should

Table 9.6. Hyperconcentrated floods without 'ripping up the bottom' phenomenon at Longmen.

Year	Time	S (kg/m³)	UJ (cm/s)	ω (cm/s)	τ_c (g/cm²)	τ_B (g/cm²)	If $UJ > \omega$	If $\tau_B \gg \tau_c$	If solid bed
1966	July 27-29	434	0.32	0.049	0.042	0.533	Yes	Yes	Yes
1977	Aug. 9-11	547	0.15	0.039	0.047	0.23	Yes	Yes	Yes

Table 9.7. Criteria for 'ripping up the bottom' phenomenon.

Author	Criterion for erosion	Criterion for suspension	Other criterion
Wan & Sheng (1975)	Pieces of bed can be eroded $\dfrac{\gamma_m}{\gamma - \gamma_m} > 0.01$ m for Longmen, and > 0.0025 m for Lington		$S > 520 \,(\text{kg}/\text{m}^3)$
Wuhan Institute of Hydraulic and Electric Engineering	Pieces of bed can be eroded $\dfrac{U^2/2g}{C_1 C_2 (At)^{2/3}} > \dfrac{\gamma - \gamma_m}{\gamma_m}$	Hyperconcentration	
Wang & Gu (1982)	The shear acting on the bed is much larger than the threshold shear: $\tau_B \gg \tau_c$	All the suspended load are wash load: $\omega < UJ$	Solid bed: $C_{6b} > C_m$
Miao & Fang (1983)		Hyperconcentration $(S > 400 \,\text{kg}/\text{m}^3)$ lasts more than 10 hrs	Large discharge $(Q > 6000 \,\text{m}^3/\text{s})$ lasts more than 5-6 hrs
Wang (1986)		The eroded particles turn to be wash load: $\omega_b / \beta k u_* < 0.06$	

be some fine particles in bed material and then the erosion of bed would take the form of pieces. But this idea is not reflected in his criteria.

All the aforementioned criteria can be summarized in Table 9.7.

Recently Wan & Song (1991) further analysed data of Longmen Gauging Station and suggested the following criteria for 'ripping up the bottom'.

1. The flow intensity is high enough to lift pieces of bed. Based on field data, this condition can be written as:

$$\theta = \frac{\gamma H J}{\gamma' - \gamma} > 0.01 \tag{9.12}$$

The definition of symbols H, J in Equation (9.12) has been given. Here the unit weight γ' of the saturated bed deposit is taken as $1.87 \,\text{t}/\text{m}^3$.

2. The concentration is so high that eroded sediment can be easily carried away by the flow.

Suspension energy $S\omega$ can be regarded as an index, reflecting the load to the flow. Here S is the concentration, in kg/m^3 and ω is the gross fall velocity of sediment particles at concentration S. ω decreases with increasing concentration S, therefore the product $S\omega$ does not monotonously increase with increasing S. For non-uniform sediment suspension energy should be written as $\Sigma S_i \omega_i$. Here S_i is the concentration of i-group sediment particles and ω_i is the corresponding gross fall velocity.

Based on measured field data, $\Sigma S_i \omega_i - S(\Sigma S_i)$ curves were plotted for several

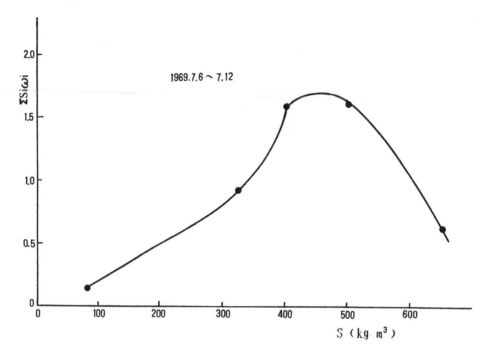

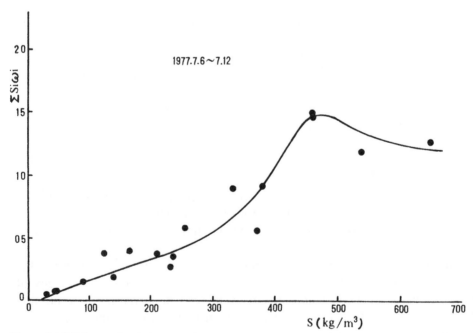

Figure 9.15. $\Sigma S_i \omega_i - S$.

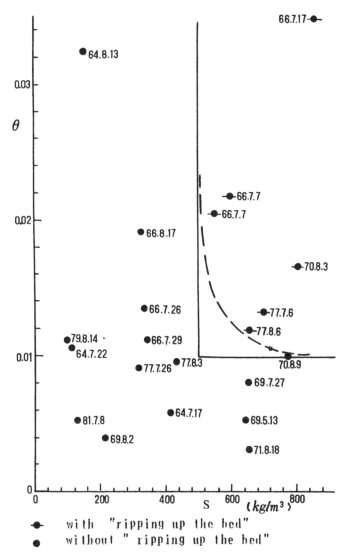

◆—	with "ripping up the bed"
●	without "ripping up the bed"

Figure 9.16. $\Sigma S_i \omega_i - S$.

floods. It is found that for floods with 'ripping up the bottom' phenomenon $\Sigma S_i \omega_i$ − S curves always have a convex outline with a maximum value of $\Sigma S_i \omega_i$, located around concentration $S = 500$, as shown in Figure 9.15.

For floods without 'ripping up the bottom' phenomenon $\Sigma S_i \omega_i$ values always monotonously increase with increasing concentration. Such tendency of $\Sigma S_i \omega_i$ − S reveals the following fact: as soon as the concentration of some floods exceeds a certain critical value, for instance, 500 kg/m³, further adding sediment into the flow does not increase the suspension energy. It means that the eroded sediment can be carried away and the erosion continues. That is just the 'ripping up the bottom' phenomenon.

Combining the aforementioned two conditions, we plotted $\theta - S$ figure for each flood, see Figure 9.16. In the figure, floods with 'ripping up the bottom' phenomenon and those without 'ripping up the bottom' phenomenon are marked by different symbols. Two kinds of dots are distributed in different regions. Dots corresponding to 'ripping up the bottom' phenomenon are concentrated in region of $\theta > 0.01$ and $S > 500$ kg/m^3. It means that 'ripping up the bottom' phenomenon happens whenever the concentration is higher than 500 kg/m^3 and θ, a parameter reflecting the flow intensity, is higher than 0.01.

In the paper they also explained the reason why pieces of bed, instead of

Table 9.8. Threshold condition of individual particles in the period of floods with 'ripping up the bottom'.

Time	$\dfrac{\gamma HJ}{(\gamma_s - \gamma)D}$	τ_B (N/m^2)	$0.047 + 4.4\dfrac{\gamma HJ}{(\gamma_s - \gamma)D}$	Yes or no*
1964.7.6-7.8	11.89	1.44	20.87	No
1966.7.17-7.20	52.0	4.53	71.54	No
1970.8.2-8.5	22.73	1.82	25.63	No
	21.29	2.53	38.77	No
	19.59	1.92	27.67	No
1977.7.6-7.12	19.18	2.67	38.55	No
1977.8.5.-8.9	18.81	1.87	28.70	No
	12.83	1.46	21.47	No

*'No' means that individual particles cannot be moved.

Table 9.9. Threshold condition of individual particles in the period of floods without 'ripping up the bottom'.

Time	$\dfrac{\gamma HJ}{(\gamma_s - \gamma)D}$	τ_B (N/m^2)	$0.047 + 4.4\dfrac{\gamma HJ}{(\gamma_s - \gamma)D}$	Yes or no*	Depth of erosion (m)**
64.7.16-7.18	13.06	0.625	7.92	Yes	−2.04
7.21-7.23	9.29	0.090	1.1	Yes	−1.48
8.13-8.14	9.66	0.50	6.29	Yes	−3.53
66.7.26-7.28	24.85	0.61	7.5	Yes	−1.70
7.28-8.1	20.5	0.60	7.4	Yes	−1.0
8.16-8.17	35.11	0.60	7.35	Yes	−18.5
66.5.12-5.15	9.44	11.0	155	No	+2.8
7.28-7.29	8.4	17.0	250	No	−2.4
7.31-8.2	4.45	0.20	2.52	Yes	−2.96
70.8.9-8.11	13.26	2.5	37.64	No	+1.20
71.7.24-7.26	6.05	0.337	4.26	Yes	−3.0
8.17-8.19	5.14	0.90	12.76	No	=0
77.8.3-8.4	8.8	0.56	7.6	Yes	−0.89
8.11-8.15	23	0.025	0.32	Yes	=0
81.7.7-7.9	5.5	0.278	3.43	Yes	=0

*'No' means that individual particles cannot be moved; 'Yes' means that individual particles can be moved. ** + means deposition; − means erosion.

individual particles, are eroded. The threshold shear stress (in dimensionless form) for a particles in hyperconcentrated flow should be as follows:

$$\frac{\gamma H J}{(\gamma_s - \gamma) D} = 0.047 + \frac{\tau_B}{(\gamma_s - \gamma) D} \qquad (9.13)$$

Yield stress τ_B increases rapidly with concentration increasing. So a particle in hyperconcentrated flow is not easily eroded if the concentration is high enough. Wan and Song checked the threshold condition for a series of floods. Results are listed in Table 9.8.

In the period of $\theta > 0.01$ for floods with 'ripping up the bed' phenomenon, $\gamma H J / (\gamma_s - \gamma) D$ is always less than $0.047 + \tau_B / (\gamma_s - \gamma) D$. It means that under such condition even though pieces of bed can be lifted, individual particles remain stationary.

For floods without 'ripping up the bottom' phenomenon, Equation (9.13) is fulfilled. It means that individual particles can be moved and the bed is eroded in ordinary way. It is also found in Table 9.9 that the bed was silted when Equation (9.13) was not fulfilled, that is, the left hand side was smaller.

REFERENCES

Engelund, F. & Z. Wan 1984. Instability of hyperconcentrated flow. *J. Hyd. Engin., Proc. ASCE*, Vol. 110(HY3): 219-232.

Janda, R.J. & D.F. Meyer 1985. Channel morphology changes causes by debris flow, hyperconcentrated sediment-laden streamflow, Toutle River, Mount St. Helens, Washington. *Proc. of International Workshop on Flow at Hyperconcentrations of sediment*, pp.3-6-11.

Miao, F. & Z. Fang 1984. Discussion on the mechanism of 'ripping up the bottom' phenomenon (in Chinese). *People's Yellow River*, 1984, No.1: 25-29.

Qi P. & Y. Zhao 1985. The characteristics of sediment transport and problems of bed formation by flood with hyperconcentration of sediment in the Yellow River (in Chinese). *Proc. of International Workshop on Flow at Hyperconcentrations of Sediment*, pp. 3-3-17.

Qi, P. & Y. Zhao 1987. Characteristics of hyperconcentrated floods passing through Sanmenxia Reservoir in 1977 (in Chinese). *Report of Institute of Hydraulic Research, Yellow River Conservancy Commission*.

Qi, P., Y. Zhao & Z. Fan 1984. The propagation of hyperconcentrated floods in 1977 and relative fluvial processes in the Lower Yellow River (in Chinese). *Report of Institute of Hydraulic Research, Yellow River Conservancy Commission*.

Qian, N. (Ning Chien) 1980, Preliminary investigation on the mechanism of motion of hyperconcentrated flow in the north-western China. *Selected papers on Sediment Problem of Yellow River*, Vol. 4: 244-267.

Qian, N. (Ning Chien), Z. Wan & Y. Qian 1979. The flow with heavy sediment concentration in the Yellow River basin (in Chinese). *J.of Tsinghua University*, Vol. 19(2):1-17.

Wan, Z. 1982. Bed material movement in hyperconcentrated flow. *Series Paper No.31*,

Institute of Hydrodynamics and Hydraulic Engineering, Technical University of Denmark, pp.79.

Wan, Z. & S. Sheng 1978. Hyperconcentrated flow phenomena in the main stem and tributaries of the Yellow River (in Chinese). *Selected papers on Sediment Problem of Yellow River*, Vol. 1: 141-158.

Wan, Z. & T. Song 1991. Analysis on 'ripping up the bottom' phenomenon (in Chinese). *J. of Sediment Research*, Vol. 3: 20-27.

Wan, Z., Y. Qian & N. Qian (Ning Chien) 1978. Hyperconcentrated flow on the Yellow River (in Chinese). *Report of Institute of Hydraulic Research, Yellow River conservancy Commission.*

Wan, Z., Y. Qian, W. Yang & W. Zhao 1979. An experimental study on Hyperconcentrated flow (in Chinese). *People's Yellow River*, No.1: 53-65.

Wang, S. & Y. Gu 1982. A preliminary study on the bottom scouring of the Yellow River (in Chinese). *J. of Sediment Research*, No.2: 36-44.

Wang, Z. 1986. A study on the mechanism of suspended load motion (in Chinese). *J. of Hydraulic Engineering*, No.7: 11-20.

Wuhan Institute of Hydraulic and Electric Engineering 1982. *River Engineering*, Water Conservancy Press.

CHAPTER 10

Hyperconcentrated density current

10.1 GENERAL DESCRIPTION

Hyperconcentrated density current happens quite often in reservoirs located in loess plateau. Due to the large density difference between the current and clear water, hyperconcentrated density current moves at much higher velocity than low concentration density current.

Because hyperconcentrated fluid has yield stress and large viscosity sediment particles in it do not settle or settle very slowly. Consequently, no obvious sorting phenomenon happens and high concentration can be maintained while a hyperconcentrated density current passes through a reservoir. When a hyperconcentrated density current reaches the dam and if the sluicing discharge is less than that of the incoming flow, the turbid water will accumulate there and a turbid water subreservoir will be formed in front of the dam. That is, accumulated underneath the clear water, muddy water forms a reservoir. Sediment particles in turbid water settle extremely slowly and the turbid water keeps its fluidity for a rather long time. The turbid water can still be released even if the incoming flood has ceased. In many cases, particularly in some small tributaries the duration of releasing turbid water can be much longer than the time period of a flash flood. It is quite different from a low concentration density current.

As a summary, compared with those of a low concentration density current, the velocity and concentration of a hyperconcentrated density current can be much higher, and the duration of releasing sediment can be much longer. As a combined effect of these factors, the releasing ratio of a flood by a hyperconcentrated density current can be much higher than that by a low concentration density current. In other words, sluicing sediment by hyperconcentrated density current is one of the effective ways in reservoir operation.

As an example showing the aforementioned features, several issues of hyperconcentrated density current recorded in Hengshan Reservoir, Shanxi Province (Guo et al., 1980) are listed in Table 10.1.

Table 10.1. Hyperconcentrated density current in Hengshan Reservoir.

Date	Inflow				Outflow			
	T_i (hr)	S_{max} (kg/m^3)	D_{50} (mm)	P (%)	T_o (hr)	S_{max} (kg/m^3)	D_{50} (mm)	P (%)
June 25, 1973	12.5	469			206	1220		
July 7, 1975	20.5	649			59	94		
July 23, 1976	12.0	407	0.019	14.2	20	851		
Aug. 15, 1980	112.6	462	0.015	20.3	164.5	1010	0.023	28.9
July 24, 1981	3.4	835	0.04	1.5	21.9	1070	0.056	10

Remark: P = percentage of particles finer than 0.01 mm

10.2 THE CONDITION OF PLUNGING

An open channel flow at low concentration submerges and turns to be a density current at the point where the following criterion is fulfilled (Fan, 1980):

$$\frac{q^2}{\frac{\Delta\gamma}{\gamma'}gH^3} = 0.6 \tag{10.1}$$

or

$$\mathrm{Fr}' = \frac{V}{\sqrt{\frac{\Delta\gamma}{\gamma'}gH}} = 0.78 \tag{10.2}$$

in which q is unit discharge, V and H the average velocity and the average depth, respectively, g gravity, γ' the specific weight of the fluid in lower layer, that is, the sediment-laden flow in our case, $\Delta\gamma$ the difference of specific weight between the lower layer and the upper layer, Fr' the modified Froud number.

At the plunging section intensive shear exists at the interface between the upper clear water and the lower sediment-laden flow. 'Boiling' phenomenon caused by the intensive violence at the interface can be observed on the water surface. In plane the sediment-laden flow submerges with a tongue-like front, which is visualized by the gathered odds and ends, see Figure 10.1a.

As the sediment concentration of the lower layer increases, vortex attenuates and some changes are associated. Based on phenomenon observed in flume experiments, 'plunging' can be classified into three categories (Cao et al., 1985).

Provided either the concentration of sediment-laden flow is not very high or the discharge of a hyperconcentrated flow is large, plunging phenomenon has no great difference from that of a low concenration density current described before. It is the type *A* of plunging.

For a hyperconcentrated flow of low intensity the flow pattern of plunging is quite different. Hyperconcentrated flow submerges smoothly and there is a distinct interface between the turbid water and the clear water. No 'boiling'

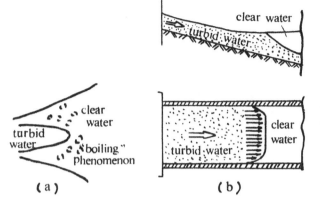

Figure 10.1. Flow patterns
near plunging section.
(a) common density current;
(b) hyperconcentrated density
current.

phenomenon and no intensive vortex is observed. Along the transverse direction
flow velocity uniformly distributes,and there is no velocity gradient execpt
closely by the walls, see Figure 10.1b (Jiao,1987). It is the type *C* of plunging.

Between these two types there is a transitional type *B*. A slight rolling surface
occurs at the plunging section and an embryo point of inflection appears downstream of the plunging section. Type *B* occurs when the concentration and
discharge have median values.

Flow pattern of plunging can be classified according to a modified Reynolds
number Re_1,which is similar to that suggested in Chapter 6, as follows:

$$Re_1 = \frac{4\gamma_m HU}{g\left(\eta + \dfrac{H\tau_B}{2U}\right)}$$

(10.3)

in which *H* and *U* are the depth and the average velocity at the plunging section,
γ_m, η and τ_B are the specific weight, rigidity and yield stress of the turbid water.

The criterion is as follow:
- $Re_1 > 5000$ type *A*   turbulent flow;
- $300 < Re_1 < 5000$ type *B*   transitional flow;
- $Re_1 < 300$ type *C*   laminar flow.

Based on a set of flume experiments, as well as some field data, Cao (1985)
established a relationship between Fr′ and $\Delta\gamma/\gamma'$ as a criterion of plunging, that is,
forming a density current, see Figure 10.2.

The following conclusion can be deduced from the figure:

1. Under the condition of small $\Delta\gamma/\gamma'$, Equation (10.2), that is, Fr′ = 0.78 is
valid as a criterion of plunging.

2. As $\Delta\gamma/\gamma'$ increases, Fr′ obviously decreases and deviates from the constant
0.78.

3. As $\Delta\gamma/\gamma'$ reaches 0.2-0.25, Fr′ rapidly drops down. It is difficult to calculate

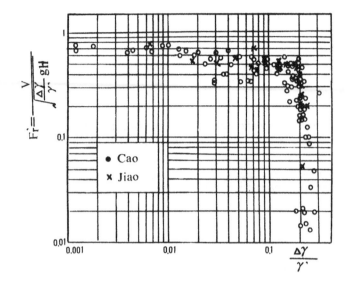

Figure 10.2. Fr' – $\Delta\gamma/\gamma'$ (combined).

the depth H at the plunging section on the basis of this figure.

As we will discuss it later, large $\Delta\gamma/\gamma'$ corresponds to high concentration. In the case of $\Delta\gamma/\gamma' > 0.2\text{-}0.25$, the flow is a hyperconcentrated laminar one, and Cao et al. suggested that the depth at the plunging section can be determined as follow:

$$H = \frac{K_p \tau_B}{64\Delta\gamma J} \tag{10.4}$$

in which $K_p = 150$.

A figure similar to Figure 10.2 was plotted by Jiao (1987). In Jiao's figure the abscissa in Figure 10.2 is replaced by the volumetric concentration S_v.

The specific weight of turbid water γ' and the difference of specific weight $\Delta\gamma$ can be related to volumetric concentration S_v as follows:

$$\gamma' = \gamma_0 + (\gamma_s - \gamma_0)\,S_v \tag{10.5}$$

$$\Delta\gamma = \gamma' - \gamma_0 = (\gamma_s - \gamma_0)\,S_v \tag{10.6}$$

in which γ_s and γ_0 are the specific weight of sediment particles and that of clear water, respectively.

Since the specific weight of natural sediment particles varies in a narrow range, for instance, 2.6-2.7 g/cm^3, we can take 2.65 g/cm^3 as an average, and $s = \gamma_s / \gamma_0 = 2.65$. The abscissa of Figure 10.2 $\Delta\gamma/\gamma'$ can be directly related to the concentration S_v as follows:

$$\frac{\Delta\gamma}{\gamma'} = \frac{(s-1)\,S_v}{1+(s-1)S_v} = \frac{1.65\,S_v}{1+1.65\,S_v} \tag{10.7}$$

$\Delta\gamma/\gamma'$ is a monotropic function of S_v. It means that Cao's figure and Jiao's figure are interchangeable. Here Jiao's figure is transformed according to Equation (10.7) and dots from his figure are added into Figure 10.2 and marked by special symbols. It can be seen from the figure that all the dots follow a common tendency and no obvious deviation can be detected.

Further developing his deduction, Jiao (1989) thinks viscous force plays an important role in the processes of forming a hyperconcentrated density current. The effect of the viscous force can by considered by introducing a modified Reynolds number:

$$\text{Re}_m = \frac{4\gamma'UH}{\frac{\Delta\gamma}{\gamma'}g\left(\eta + \frac{\tau_B H}{2U}\right)} \tag{10.8}$$

Using the same data used in $\text{Fr}' - S_v$ figure, he developed an empirical relationship between $(\text{Fr}' \, \text{Re}_m)^{0.5}$ and S_v as a criterion of plunging.

It should be pointed out that the modified Reynolds number used here is different from the commonly used one. Here an additional term $\Delta\gamma/\gamma'$ is introduced in denominator. The reason for doing so is not clear.

10.3 RESISTANCE OF A HYPERCONCENTRATED DENSITY CURRENT

For different flow patterns flow resistance obeys different laws. (Cao, 1983).

For a turbulent density current, the modefied Reynolds number Re_1 is larger than 8000, the friction factor λ is a constant, taken as 0.025.

Here

$$\lambda = \frac{8gRJ}{\frac{\Delta\gamma}{\gamma'}U^2} \tag{10.9}$$

in which H and U the depth and the mean velocity of the density current, τ_B and η the yield stress and the rigidity of the turbid water, J slope, R hydraulic radius.

Most hyperconcentrated density current behaves as a laminar flow. For a laminar density current, the friction factor λ is proportional to the reciprocal of Re_1, see Figure 10.3.

$$\lambda = \frac{K}{\text{Re}_1} \tag{10.10}$$

The coefficient K is no longer a constant, but varies in the range of 96-384. These two figures correspond to the coefficient in the case of an open channel flow and that in the case of a flow between two parallel plates, respectively. It means that in the case of a density current besides the resistance along the bottom a resistance exists along the interface between the upper layer and the lower layer and the

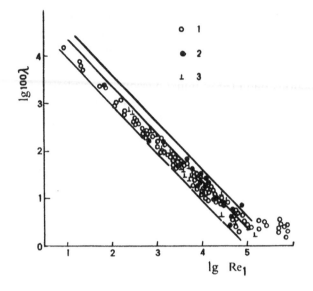

Figure 10.3. λ – Re_1. 1. Experimental data 4-600 kg/m³; 2. Hongshan 87-207 kg/m³; 3. Bajiazui 66-615 kg/m³.

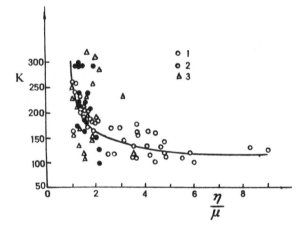

Figure 10.4. K – η/μ. 1. Experimental data 4-600 kg/m³; 2. Hongshan 87-207 kg/m³; 3. Bajiazui 66-615 kg/m³.

resistance is smaller than the resistance along a fixed upper plate, which exists in the case of a flow between two parallel plates.

Cao assumes that K depends on fluid properties of both upper layer and lower layer. Based on experiment data, she established a correlationship between K and η/μ as follows, see Figure 10.4.

$$K = 96 + [166 \left(\frac{\eta}{\mu}\right)^{-1.3} + 3]$$

(10.11)

in which μ is the viscosity of upper clear water.

Zhao & Zhou (1987) further discussed the resistance of a hyperconcentrated density current. Coefficient K can be considered as an index reflecting the ratio of

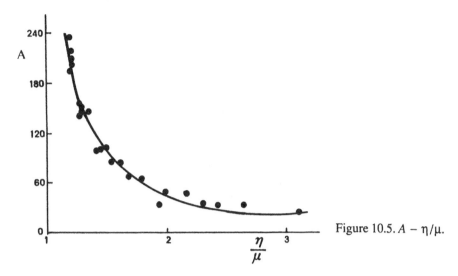

Figure 10.5. $A - \eta/\mu$.

the bottom resistance to the interface resistance, which depends on not only the fluid properties, but also the flow conditions. Richardson number, which is defined as Equation (10.12), is an important parameter in density current.

$$\mathrm{Ri} = \frac{\dfrac{\Delta\gamma}{\gamma'}\dfrac{g}{H}}{\left(\dfrac{U}{H}\right)^2} = \frac{g'H}{U^2} = 1/\mathrm{Fr}'^2 \tag{10.12}$$

Here $g' = (\Delta\gamma/\gamma') g$. Richardson number, which is the reciprocal of the square of the modified Froud number, reflects the portion of energy, spent on overcoming the density gradient. Therefore it can be imagined that Ri plays a role in the determination of flow resistance. Zhao and Zhou carried out a set of experiments on density current in a tilting flume ($22 \times 0.5 \times 0.6$ m). The median diameter (D_{50}) of sediment they used is 0.022 mm. Based on experiment data they established a correlationship between $A = 1570\, K\mathrm{Ri}^{-0.5}\, g\mu/\gamma'q$ and η/μ as shown in Figure 10.5. Here q is the unit discharge.

The correlationship can be written as an emperical formula:

$$K = 0.188 \left(\frac{\eta}{\mu}\right)^{-2.45} \mathrm{Ri}^{0.5}\frac{\gamma q}{g\mu} \tag{10.13}$$

Substituting Equation (10.13) into Equation (10.10) and simplifying it, one obtained:

$$\lambda = 0.188 \left(\frac{\eta}{\mu}\right)^{-2.45} \mathrm{Ri}^{0.5}\frac{\gamma'q}{g\mu}\mathrm{Re}_1 \tag{10.14}$$

or

$$\lambda = \frac{0.047}{\mathrm{Fr}'} \left[\left(\frac{\eta}{\mu}\right)^{-1.45} + \left(\frac{\eta}{\mu}\right)^{-2.45} \frac{\tau_B R}{2U\mu} \right] \tag{10.15}$$

10.4 THE CONTINUOUS MOTION OF A HYPERCONCENTRATED DENSITY CURRENT

Conditions for keeping a continuous motion of a density current have been studied by Fan (1959). Based on data from Guanting and other reservoirs, he concluded as follows:once the flood entering the reservoir vanishes, the density current induced by the flood ceases moving, and the sediment carried by the density current will settle and form an appreciable deposition. Therefore, the condition for keeping a continuous motion of a density current depends on the time-period of a flood. And the condition for releasing sediment from a reservoir by density current is that the duration of a flood must be longer than the time required for the density current travelling from the plunging section to the front of the dam.

The situation of a hyperconcentrated density current is quite different. Due to the much larger density difference, the effect of gravity on a hyperconcentrated density current is more intensive. It means that the velocity of a hyperconcentrated density current could be much higher. Besides, even if the flood entering the reservoir vanishes, a hyperconcentrated density current may keep moving downstreamward under the gravity. Due to the yield stress and large rigidity of the hyperconcentrated fluid coarse particles in a hyperconcentrated density current settle very slowly. Therefore sorting phenomenon does not exist and no sediment particle deposits along its course.

As an example, grain composition of samples taken at different places of Bajiazui Reservoir during a hyperconcentrated density current are shown in Table 10.2. For Bajiazui Reservoir incoming flow comes from two tributaries, on which Yaoxinzhong Gauging Station and Taibeiliang Gauging Station are located, respectively. And Bajiazui Gauging station is located downstream from the dam.

It can be seen from the table that throughout the whole reservoir, that is, from its inlets to its outlet sediment grain composition of a hyperconcentrated density current kept nearly unchanged. In other words, coarse particles did not settle and no sorting phenomenon happened. Table 10.3 presents another example.

Correspondingly, grain composition of bed material do not change too much along the reservoir. One example of sampling taken in January, 1972 is listed in Table 10.4.

Once a hyperconcentrated density current reaches the dam and the sluicing discharge is less than the incoming discharge, turbid water is accumulated in front of the dam and a subreservoir consisting of turbid water is formed underneath the clear water. An interface between the clear water and turbid water is distinct. Due

Table 10.2. Variation of grain composition of a hyperconcentrated density current (May 26-27, 1984).

Place	Distance to the dam (km)	Time	Concentration S (kg/m^3)	Percentage of sediment particles finer than (mm) 0.005	0.01	0.025	0.05	0.1	0.25	0.5	1.0
Yaoxinzhong (inlet G.S.)	30.77	May 26.2:00	380	22.7	31.4	53.2	82.0	99.5	100		
Teibeiliang (inlet G.S.)	34.9	May 26.1:36	714	20.4	28.6	49.0	79.4	99.2	99.8	100	
In reservoir	8.47	May 26.10:00	467	20.3	31.0	56.7	90.6	99.8	100		
	7.11	10:00	461	23.2	31.8	60.2	89.9	99.8	100		
	6.13	11:00	468	23.4	33.2	60.4	96.0	99.8	100		
		12:00	455	24.5	33.9	63.1	96.0	99.8	100		
	2.05	14:00	431	23.1	33.0	60.5	92.3	99.9	100		
	0.54	14:00	351	27.0	37.6	68.8	89.2	99.9	100		
Bajiazui (outlet G.S.)		May 26.05:30	331	19.3	25.5	51.6	86.1	99.7		99.9	100
		08:00	655	15.5	18.6	44.5	83.8	99.3	100		
		16:00	380	27.7	35.2	64.5	90.7	99.7	100		
		May 27.06:00	103	26.6	36.6	66.1	90.5	99.9	100		

Table 10.3. Variation of grain composition of a hyperconcentrated density current (September 7, 1983).

Place	Distance to the dam (km)	Time	Percentage of sediment particles finer (%) 0.005	0.01	0.025	0.05	0.1	0.25	0.5	1.0	D_{50} (mm)	D_{90} (mm)
Yaoxinzhong (inlet G.S.)	30.77	Sept.7. 5:00	18.0	26.4	49.7	83.7	95.6	98.0	99.4	99.8		
		Sept.7	17.7	26.4	49.0	81.0	97.6	99.4	100		0.026	
Teibeiliang (inlet G.S.)	34.9	Sept.7. 5:00	19.6	27.7	53.1	85.0	99.4	99.8	100			
		Sept.7	20.0	29.9	52.8	83.4	99.4	100			0.023	
In reservoir		Sept.7.11:00	19.5	28.9	48.4	80.2	99.7	100			0.026	0.062
			19.4	30.3	55.1	82.4	99.7	100			0.022	0.057
		Sept.7.13:00	18.3	29.2	55.6	88.9	99.7	100			0.022	0.050
			21.2	30.7	54.3	85.6	99.7	100			0.022	0.054
	2.05	Sept.7.14:00	20.6	29.8	57.1	88.1	99.8	100			0.021	0.053
			20.1	29.0	53.5	88.6	99.7	100			0.023	0.053
	0.54	Sept.7.10:00	19.7	29.1	54.6	87.0	99.7	100			0.022	0.054
			18.7	26.6	52.3	85.3	99.6	100			0.023	0.056
Bajiazui (outlet G.S.)		Sept.7.10:00	18.1	25.2	49.7	79.9	98.4	99.8	100			
		Sept.7.20:00	20.7	29.8	57.2	89.4	99.7	100				
			18.1	26.5	49.9	82.7	99.3		100		0.025	0.059

Table 10.4. Longitudinal D_{50} variation of deposits along a reservoir.

Distance to the dam (km)											
0.0	0.54	2.05	5.07	6.13	7.11	8.47	11.11	12.18	13.33	15.63	16.62
D_{50}: 0.0425	0.0425	0.0463	0.0397	0.0383	0.0335	0.049	0.0374	0.0411	0.043	0.045	0.0439
(mm)											

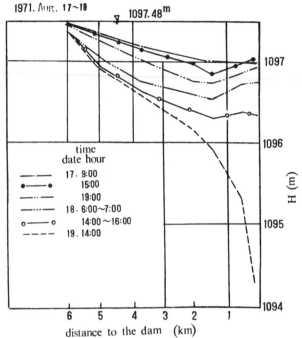

1971. Aug. 17~19   ▽ 1097.48ᵐ

time
date hour
17. 9:00
15:00
19:00
18. 6:00~7:00
14:00~16:00
19. 14:00

distance to the dam   (km)

Figure 10.6. Settling of the interface between muddy water and clear water.

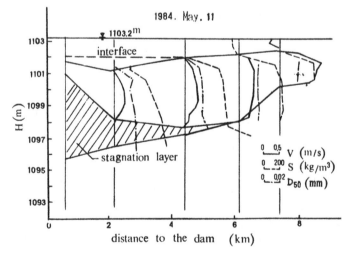

1984. May. 11

▽ 1103.2ᵐ

interface

stagnation layer

0 0.5 V (m/s)
0 200 S (kg/m³)
0 0.02 D₅₀ (mm)

distance to the dam   (km)

Figure 10.7. Vertical profiles of velocity, concentration and D_{50} in a density current.

to the high concentration sediment particles settle extremely slowly and the interface lowers at a very low speed. Figure 10.6 is one of the examples.

The turbid water in subreservoir maintains high concentration and keeps its fluidity for a rather long time. Only near the bottom there is a stagnation layer, which keeps stationary and has extremely high concentration, e.g. 1000 Kg/m³, see Figure 10.7.

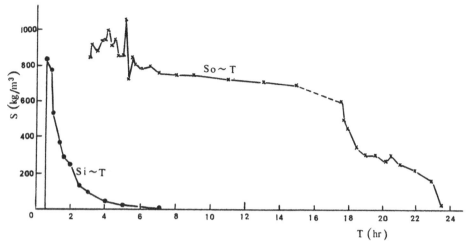

Figure 10.8. Hydrograph of inflow and outflow sediment concentration during releasing muddy water from underlying subreservoir. S_o = outflow sediment concentration; S_i = inflow sediment concentration.

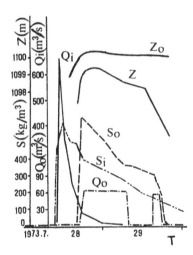

Figure 10.9. Hydrograph of incoming and outcoming discharge and concentration. Z_0, Z = clear water level and the level of interface. Q_i, Q_o = incoming and outcoming discharge. S_i, S_o = incoming and outcoming concentration.

Most muddy water in the subreservoir can still be released from the reservoir only if the bottom sluicing gate is opened, no matter whether the incoming flood has vanished or not. As a result, the duration of releasing muddy water might be much longer than the duration of an incoming flood. And in some cases the concentration of released turbid water can be even higher than that of the incoming flood due to the settling and the condensation of the turbid water. Figure 10.8 in Hengshan Reservoir is a remarkable example, showing the much longer time of releasing turbid water and also the higher concentration of the outflow. Figure 10.9 is another example, happening in Bajiazui Reservoir. The settling of the interface can also be detected from the figure.

Compared with those of a density current at low concentration, characteristics of a hyperconcentrated density current can be summarized as follows:

1. Higher velocity results from large density difference.

2. Less or no deposition occurs along its course due to the high viscosity of the turbid water. Correspondingly, no sorting phenomenon happens along its course. Sediment size at different places of the reservoir changes very little.

3. Longer duration of releasing sediment results from the fluidity, which the turbid water in a subreservoir can keep for a long time. In some cases the concentration of released turbid water can be higher than that of the incoming flow.

4. As a result of aforementioned factors, much more sediment can be released from the reservoir by hyperconcentrated density current.

10.5 HYPERCONCENTRATED DENSITY CURRENT IN RIVERS
   (WAN & NIU, 1989)

As mentioned above, a hyperconcentrated flow easily forms a submerged density current in reservoir due to its large density and the corresponding large difference in density between it and clear water. Also due to its large density a hyperconcentrated flow might form a density current in rivers at some confluences in the Yellow River Basin. Like 'clogging' and 'ripping up the bottom', density current in rivers is also an unusual phenomenon associated with hyperconcentrated flow, which has not been observed and recorded anywhere else.

Hyperconcentrated density current in rivers has been carefully observed and measured at Tongguan Gauging Station, which is located right downstream from the confluence of the Yellow River and its tributary Weihe River, see Figure 10.10. Hydrologists at the station found whenever hyperconcentrated flood coming from Weihe River met the relatively clear water flow coming from the Yellow River, abrupt change of vertical concentration distribution and submerging hyperconcentrated lower layer could always be observed at Tongguan Gauging Station. In the meantime, a mass of debris, driftwood, and leaves accumulating on the water surface along the front of plunging near the confluence clearly shows the plunging of Weihe River turbid flow. This phenomenon is quite similar to that associated with the plunging of a density current in a reservoir.

It enlightened persons on considering the problem: 'if there is density current like that in reservoir?' Careful measurement and analysis led to a positive answer.

As an example, hydrographs of discharge Q and sediment concentration S at an upstream Gauging Station of the Yellow River and those of the Weihe River are plotted in Figure 10.11. The propagation time has been considered when the figure is plotted.

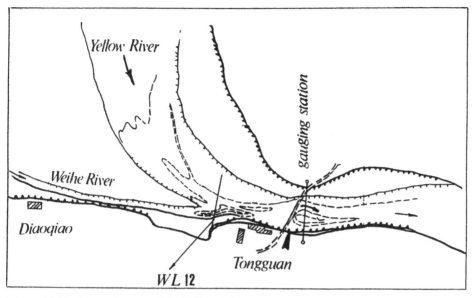

Figure 10.10. the confluence of the Yellow River and its tributary Weihe River.

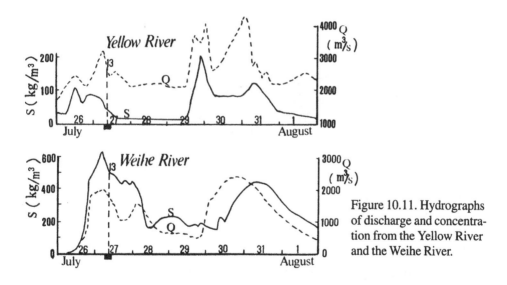

Figure 10.11. Hydrographs of discharge and concentration from the Yellow River and the Weihe River.

Measured velocity and concentration profiles and concentration distribution along the cross section at Tongguan Gauging Station are shown in Figure 10.12. The measurement was taken at the moment, which was marked by black blocks in Figure 10.11, when the sediment concentration of the flow from Weihe River was high and that from the Yellow River was low. In Figure 10.12 u and S are point velocity and concentration, respectively. Symbol with a bar means the average value along a vertical. H is the water depth.

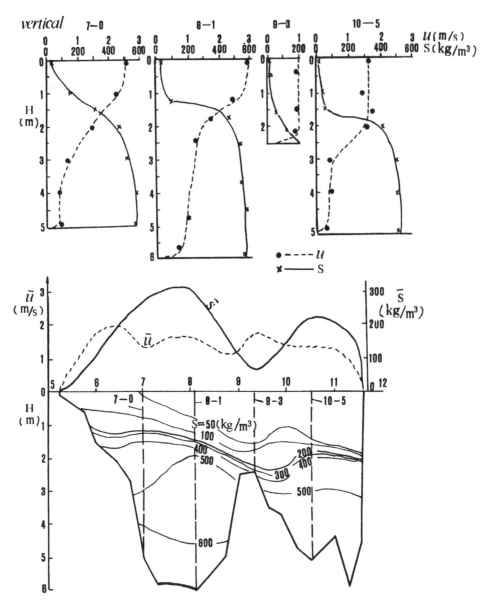

Figure 10.12. The velocity and concentration distribution at Tongguan Gauging Station.

It is obvious that there was a lower layer with high concentration. With an abrupt change the concentration of the upper layer was low. At that time it was observed that a mass of debris accumulated and formed an oblique line by the right bank of the Weihe River near the confluence of the two rivers.

The aforementioned situation is a typical one. Such phenomenon has been repeatedly measured and observed whenever hyperconcentrated flood comes

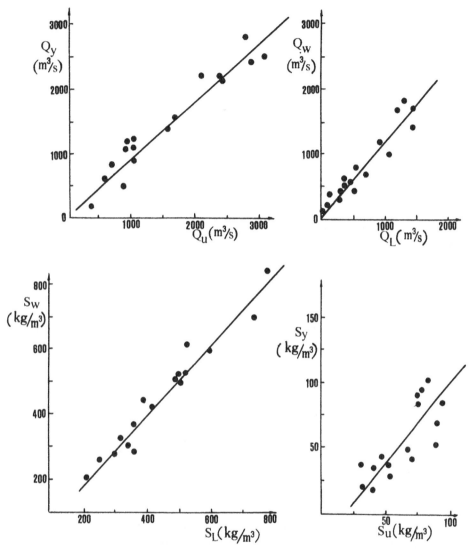

Figure 10.13. The comparison of corresponding discharges and average concentrations.

from the Weihe River and the flow from the Yellow River is relatively clear.

In order to judge if the hyperconcentrated flow from the Weihe River sub-merged underneath the flow from the Yellow River, following analysis has been made. For each time of measurement, at which an abrupt concentration change along verticals could be detected, the whole flow was divided into two parts by interface where an abrupt concentration change existed. The discharge and average concentration of the upper layer were calculated and compared with the corresponding discharge and average concentration of the flow from the Yellow River. The discharge and average concentration of the lower layer were calculated and compared with the corresponding discharge and average concentration of the flow from Weihe River. In the comparison the propagation time from the upstream guaging station has been considered and modified. The result of such comparision is shown in Figure 10.13.

In the figure Q and S stand for discharge and concentration, subscript y, w, u, l represent Yellow River, Weihe River, upper layer and lower layer, respectively. For instance, Q_y means the discharge of the flow from the Yellow River, S_L means the average concentration of the lower layer.

It can be seen from the figure that dots are distributed along straight lines with 45° and they are not scattered. It can be deduced that the hyperconcentrated flow from the Weihe River indeed submerged underneath the flow from the Yellow River and the mixing between them was not so strong. And the flow from the Yellow River ran along the upper layer. The density current like that in reservoir indeed existed.

The critical condition for a sediment-laden flow submerging underneath a reservoir can be described by Equation (10.2), as mentioned above.

Considering the foregoing critical condition for density current in reservoir, we calculated the modified Froud number at cross section WL 12, which was located just at the mouth of the Weihe River (see Figure 10.10), for each time of measurement, at which abrupt concentration change along the verticals could be detected.

$$\text{Fr}' = \frac{U}{\sqrt{(\Delta\gamma/\gamma)gH}} = \frac{Q_w}{BH\sqrt{(\Delta\gamma/\gamma)gH}} \tag{10.16}$$

in which Q_w is the discharge of the Weihe River, B, H, U are the width, average water depth and average velocity at the cross section WL12, γ is the specific weight of the hyperconcentrated flow from the Weihe River and $\Delta\gamma$ is the difference in specific weight between the flow from the Weihe River and that from the Yellow river.

Because no hydrological measurement was taken right at the WL12 cross section, the parameters appearing in Equation (10.16) were deduced indirectly. Discharge measured at a upstream Gauging Station was used as Q_w and the propagation time was considered in the calculation. The cross section WL12 was measured 2-4 times a year, and the nearest measurement result of cross section was used in the calculation of B and H.

The result of calculation is shown in Table 10.5. In most cases listed in Table 10.5 Fr$'$ < 0.78.

In fact a hyperconcentrated flow might submerge at a section upstream from WL12. The critical condition of plunging should be met at that cross section, and the average velocity and corresponding Fr$'$ at WL12 might be smaller than those at the plunging cross section. Besides, deposition or erosion might happen between two successive measurements of cross section, and the indirectly deduced U and H might deviate from the real values. And some inaccuracy may

Table 10.5. Data of density currents in rivers.

Time		Q	Discharge (m³/s) Q_u	Q_L	Q_y	Q_w	Concentration (kg/m³) S_u	S_L	S_y	S_w	Fr′
1969.7.31	15:15-18:10	3720	2800	920	2800	1200	161	508	230	510	0.723
9.2	8:00-09:37	1020	755	265	850	289	42.8	353	20	286	0.388
1970.7.27	8:30-10:48	777	389	389	250	473	62.2	303	45	279	0.72
1971.8.23	8:25-10:10	1990	910	1080	520	1000	72.0	364	85	373	0.68
1972.7.3	8:00-10:00	180	1070	310	1250	316	68.9	780	100	859	0.586
7.4	15:00-16:50	1130	1045	84.7	1150	160	30.8	744	40	625	0.373
1973.7.23	8:00-09:48	1870	1715	155	1600	213	56.0	339	35	303	0.761
8.20	8:10-10:20	1660	894	766	1100	712	89.5	584	80	592	1.27
21	15:25-16:40	1680	1140	540	900	415	49.3	307	45	328	1.45
30	8:05-19:05	3150	1580	1570	1400	1700	85.0	382	55	449	0.659
1974.7.30	8:28-10:30	1140	641	500	628	500	76.9	480	90	620	0.33
31	8:30-10:40	1110	960	150	1200	390	72.7	500	80	520	0.283
1975.7.26	23:00-01:17	4390	3150	1240	2500	1700	68.5	540	50	536	1.04
27	8:50-11:15	3560	2110	1450	2200	1450	43.3	511	35	537	0.88
29	8:35-10:45	2750	2410	344	2150	550	35.2	207	15	211	0.635
31	8:50-11:30	4190	2880	1310	2400	1850	87.5	419	70	420	1.40
8.1	8:37-10:33	2910	2370	539	2200	800	58.7	247	30	265	0.83

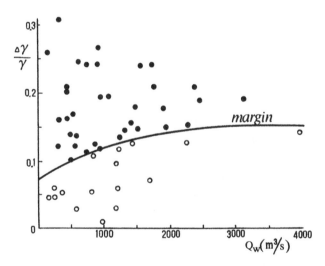

Figure 10.14. Judgement of a hyperconcentrated density current in rivers.

exist in field measurement. Considering all these factors, we can take Fr′ < 0.78 as the critical condition of forming a density current in rivers.

The modified Froud number Fr′ consists of parameters Q_w, B, H, and $\Delta\gamma/\gamma$. Provided no obvious aggradation or degradation happen, B and H vary with discharge Q_w. Therefore among these four parameters Q_w and $\Delta\gamma/\gamma$ play leading roles. Based on this idea, Figure 10.14 is plotted. Cases, in which density current happens, and cases, in which density current does not happen, are marked by different kinds of dots. A margin line can be drawn between the two kinds of dots.

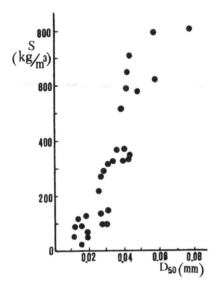

Figure 10.15. $D_{50} \sim S$ correlationship for hyper-concentrated flow (at Tongguan Gauging Station).

Therefore, whether a hyperconcentrated density current happens can be judged by these two parameters and Figure 10.14 can be used as a rule of thumb for forecast.

The density current in rivers has some characteristics.

It is similar to hyperconcentrated floods that the sediment size of a hyperconcentrated density current in rivers (lower layer) becomes coarser and coarser as its concentration increases, see Figure 10.15.

Generally speaking, sediment of density currents usually happening in reservoirs consists of grains finer than 0.025mm (Chien & Fan, 1959). It can be seen from Figure 10.15 that sediment size of hyperconcentrated density currents in rivers is much coarser. It is closely related to characteristics of hyperconcentrated flow. That is, the higher the concentration, the coarser the particles.

The concentration profiles of a density current have an abrupt change. It has been described in Figure 10.12. Besides, the sediment size of suspended load also has a synchronous rapid change along verticals. As an example, granulometric curves of samples taken along a vertical are shown in Figure 10.16. Figures by the curves denote relative water depths at which samples were taken. In the upper layer sediment size becomes coarser as water depth increases. In the lower layer sediment size keeps nearly unchanged. For more careful investigation the variation of D_{50} and D_{95} along a vertical is plotted in Figure 10.17. It can be detected that in the lower layer D_{95} increases with water depth increasing, but D_{50} keeps nearly constant.

The velocity of a hyperconcentrated density current in river usually is higher than 0.7 m/s, with a maximum of 1.4 m/s. It is much higher than that of an ordinary density current in reservoir. The later usually has a velocity of 0.2-0.4 m/s (Chien & Fan, 1959).

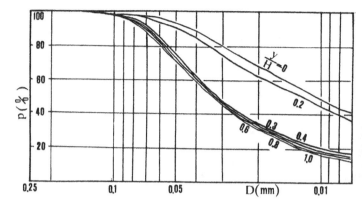

Figure 10.16. Granulometric curves of samples taken along a vertical.

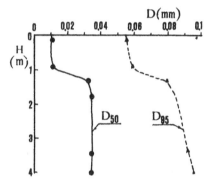

Figure 10.17. The variation of D_{95} and D_{50} along a vertical.

The lower layer of a hyperconcentrated density current in river flows in the same direction as the upper layer does. Flowing with a higher velocity, the upper layer exerts a tractive force, instead of a resistance, on the lower layer.

Carefully looking at Figure 10.12, we will find that the interface is inclined. It is because at the confluence the flow from the Weihe River turns to right and behaves as a flow in bend, see Figure 10.10. A transverse slope of the interface exists as that occurs in a bend. Compared with that of an open channel flow, the inclination of the interface of a hyperconcentrated density current is more obvious. It can be explained as follows. The magnitude of transverse slope depends on the ratio between gravity and inertia force. In case of density current in rivers, only the submerged weight plays role and the gravity is greatly reduced ($\gamma \rightarrow \Delta\gamma$). On the other hand, the inertia force increases ($\gamma \rightarrow \gamma'$). Due to the enlarged inertia force and the greatly reduced gravity, the transverse slope is greatly enlarged.

Flowing with high velocity, the hyperconcentrated density current in rivers possesses powerful sediment carrying capacity. Quite often rapid erosion, instead of serious siltation, is associated with the hyperconcentrated density current in rivers. Characteristics of erosion are as follows. At Tongguan Gauging Station if

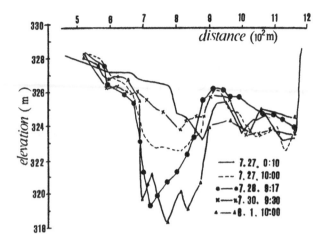

Figure 10.18. The aggradation and degradation at Tongguan Gauging Station.

no channel exists before the density current, a deep channel will form along the left bank. if there is a channel before the density current, the channel will be enlarged by the hyperconcentrated flow from the Weihe river, which turns to right at Tongguan Gauging Station. So the deep channel is always located on the left. An example is shown in Figure 10.18. Hyperconcentrated density current happened on July 27th and on August 1st, 1975. The deep channel formed by the density current of July 27th deteriorated after the density current and almost disappeared. Then the density current of August 1st eroded the bed again and moulded another deep channel.

REFERENCES

Cao, R. & J. Chen 1980. Erosion and sedimentatiion of flow at hyperconcentration in reservoirs (in Chinese). *Proc. of the First International Symposium on River Sedimentation*, p. 783-792.

Cao, R., S. Chen et al. 1983. The law of hydraulic resistance in density current with hyperconcentration (in Chinese). *Proc. of the Second International Symposium on River Sedimentation*, p. 55-56.

Cao, R., X. Ren & S. Ju 1985. Condition of formation and continuous motion of density current with hyperconcentration. *Proc. of International Workshop on Flow at Hyperconcentrations of Sediment, IRTCES, Beijing*.

Chien, N. & J. Fan 1959. *The study on density current and its application*. Press of Water Conservancy and Electric Power.

Guo, Z., B. Zhou, L. Ling & D. Li 1985. The hyperconcentrated flow and its related problems in operation at Hengshan Reservoir. *Proc. of International workshop on Flow at Hyperconcentrations of Sediment, IRTCES, Beijing*.

Jiao, E. 1987. Experimental study on hyperconcentrated density current in planning Xiaolangdi Reservoir (in Chinese). *Report of Institute of Hydraulic Research, Yellow River Conservancy Commission*.

Jiao, E. 1989. Study on sedimentation in Bajiazui Reservoir relating to hyperconcentrated flow (in Chinese). *Report of Institute of Hydraulic Research, Yellow River Conservancy Commission.*

Wan, Z. & Z. Niu 1989. Hyperconcentrated density current in rivers. *Proc. of 23th IAHR.*

Zhao, N., X. Zhou 1987. A study on resistance characteristics of hyperconcentrated density current (in Chinese). *Journal of Sediment Research*, Vol. 1: 27-34.

CHAPTER 11

Debris flow

11.1 THE CATEGORY OF DEBRIS FLOW

Debris flow is a special type of hyperconcentrated flow. Occurrence of debris flow depends highly on local geologic, topographic, meteorologic and hydrologic conditions.

Debris flow differs from an ordinary hyperconcentrated flow mainly in its gustiness, catastrophic nature and extremely high competency in carrying solid material. In hyperconcentrated flow, the flowing liquid still controls movement and deposition of sediment, and the energy consumed in the course of sediment movement comes mainly from the liquid phase. Suspended load motion and neutrally buoyant load motion play the main role in sediment transportation. In debris flow, however, laminated load motion and neutrally buoyant load motion are the dominant form of sediment motion. A considerable part of the energy supporting particles' movement is supplied by the gravitational energy of the solid phase. Denoting E_d as the energy supplied by the solid phase in unit volume and unit distance downstream, we have

$$E_d = \gamma_s S_v J \tag{11.1}$$

where J is energy slope, S_v the volume concentration of solid phase, γ_s the specific weight of particle. Plotting the data measured in hyperconcentrated flows and debris flows in Figure 11.1, Qian & Wang (1984) found that the points of debris flows and hyperconcentrated flows are distributed in different areas in the $S_v - J$ chart. The line

$$E_d = 0.01 \ (\text{g/cm}^3) \text{ or } S_v = 0.01 \ (\text{g/cm}^3) \ / \gamma_s J$$

bounds the two areas. Thus $E_d = 0.01 \ (\text{g/cm}^3)$ can be used to differentiate debris flow from normal hyperconcentrated flow. If the slope and the concentration are so high that $E_d > 0.01 \ (\text{g/cm}^3)$, the flow belongs to the category of debris flow.

Mechanical properties of debris flow depend highly on concentration and composition of solid material. Debris flows are classified into viscous, subviscous

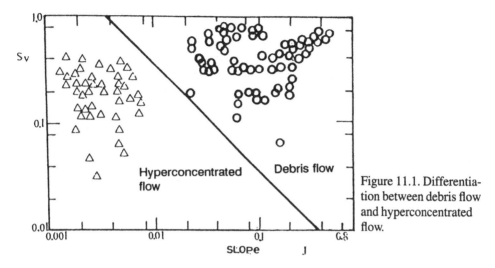

Figure 11.1. Differentiation between debris flow and hyperconcentrated flow.

Table 11.1. The dynamic characteristics of debris flows.

Type	Flow patterns	Specific weight of the mixture (g/cm³)	
		Tibet Plateau	Yunnan Province
Non-viscous debris flow	Turbulent flow with suspended load and bed load (sometimes laminated load)	1.1-1.8	1.35-1.69
Subviscous debris flow	Transitional flow with suspended load, laminated load and neutrally buoyant load	1.8-2.0	1.69-1.94
Viscous debris flow	Laminar flow with neutrally buoyant load and laminated load	>2.0	1.89-2.24

and non-viscous debris flows based on flow characteristics and the density or the specific weight of the debris-water mixtures. Table 11.1 describes the dynamic properties of the three kinds of debris flows. Because of the differences in mineral and petrological composition of solid material in different areas, the index of the specific weight of the three kinds of debris flows is different in different areas.

Viscous debris flow has large yield stress, it is laminar and often develops into intermittent flow. Subviscous debris flow has small yield stress and the flow is continuous and turbulent. Non-viscous debris flow is Newtonian and highly turbulent.

Debris flows can also be classified into mud flow, mud-rock flow and water-rock flow according to the composition of solid materials (Du et al., 1986). The mud-rock flow consists of mud and stones. Size of the solid particles ranges from clay finer than 0.001 mm to boulders of several or tens meters in diameter. Clay particles make up a certain proportion of the total amount of sediment, generally about 3-5%. The flocculent structure of clay particle has a great effect on dynamic properties and transport capacity, and endorses the flow some characteristic phenomena.

Mud flow often takes place in the loess plateau in China. Clay and sand make up most part of the solid material in mud flow and the content of clay is more than 15%. Mud flow is non-Newtonian and often develops into intermittent flow.

Water-rock flow mainly occurs in marble, dolomite, limestone and conglomerate rock areas, or partly in granite mountain areas. The solid material in water-rock flow is mainly composed of coarse sand, gravels and boulders. They move usually as laminated load.

Debris flow can also be classified into glacial debris flow and rainfall debris flow according to their genesis.

11.2 DISTRIBUTION OF DEBRIS FLOW IN CHINA

China is a country with a vast territory and most territory is mountainous area. More than 800 counties (about 40% of the total) have recorded debris flows and more than 60 towns were damaged by debris flows. Figure 11.2 shows the distribution of debris flow zone in China (Li & Luo, 1985). According to a preliminary investigation, there are more than 10 000 debris flow gullies throughout the country. Rainfall debris flow frequently occurs in Yunnan, Sichuan, and Gansu. The Xiaojing watershed in Yunnan province has a drainage area of 3220 km^2. There are 107 debris flow gullies in the area. More than 100, sometimes more than 2000 events of rainfall debris flows take place in the area and more than 2-3×10^7 tones of solid material is carried into the Xiaojiang River annually. Among these debris flow gullies the Jiangjia Gully is the most notorious. More than 10 events of debris flows take place in the gully every year and 28 events of

Figure 11.2. Distribution of debris flow in China (shadow areas are debris flow areas).

debris flows were recorded in 1965. The upstream gully bed is cut down 2-3 m and the mouth of the gully is aggraded 1.36 m, annually. It is no wonder that the Xiaojiang watershed is called a museum of debris flow (Chen, 1985).

In the middle reach of the Bailong River in Gansu Province almost all gullies are debris flow gullies. There are 10 debris flow gullies in every one kilometer along the river on average.

Glacial debris flow takes place mainly in the Qinghai-Tibet Plateau. The Guxiang gully in the plateau is a large glacial debris flow gully. Several tens glacial debris flows occur there annually. A glacial debris flow of very large scale occurred in the gully in 1953. The maximum depth of the debris flow was 40-95 m and the maximum discharge was estimated to 28 600 m^3/s. About 10 million cubic meter of solid material was transported by the debris flow (Du & Zhang, 1985).

Topography of the country varies from the Qinghai-Tibet Plateau to the Yungui Plateau and the Loess Plateau and then to the North China Plain like three huge terraces. Debris flow is distributed mainly in the transitional zones between these terraces. The glacial debris flow concentrates between the altitudes of 3000 and 6000 m, and the rainfall debris flow distributes in the areas of altitude between 700 and 3000 m. There are a few of debris flow gullies in North-East China and North China plain, too. The debris flow gullies developed in these areas mainly because of deforestation and other human activities.

Debris flow shows obvious periodicity. The occurrence of very active debris flow is 50-70 years. Overlapping on the long period are 11-year and 22-year short periods. 1960s and 1980s are two active periods of debris flow. Debris flows take places simultaneously in the Tibet plateau, Sichuan Province, Gansu, Shaan-xi, Liaoning and Jilin Provinces in 1981. Only in Sichuan Province 61 counties were hit by active debris flows.

11.3 PHENOMENOLOGICAL DESCRIPTION OF DEBRIS FLOW

11.3.1 *General aspect*

Most of debris flows are mud-rock flow. They are poorly sorted mixtures, and are often described as resembling the flow of wet concrete.

Jiang-jia Gully is a frequently-occurring debris flow gully in Yun-nan Province. The slope of the gully bed is about 0.10. Debris flow occurs many times each year. Solid material transported by debris flow has dammed the Xiaojing River 7 times since 1919. Gravels, pebbles and boulders compose 70% of the solid material and fine particles of diameter less than 2 mm are less than 30% in general.

Debris flow general occurs during or after rainstorm in summer and typically begins with torrential flood. Following erosion of the gully bed solid concentra-

Table 11.2. Dynamic indexes of the flow during development process of a debris flow in the Jiangjia Gully (1974).

No (in order)	Flow pattern	Sp. weight γ_m (g/cm)	τ_B (dyne/cm^2)	Rigidity η (Poise)	Mean diameter D (mm)	Velocity U (m/s)	Depth H (m)
1	Torrential flood	1.112	0.39	0.022	0.121		
2	Torrential flood	1.147	1.27	0.088	0.106		
3	Continuous non-viscous debris flow	1.567	15.20	0.495	1.410	3.98	0.17
4	Intermittent viscous debris flow	1.830	93.00	3.843	3.357	3.61	0.17
5	Intermittent viscous debris flow	1.995	244.60	8.76	4.973	7.50	0.45
6	Intermittent viscous debris flow	2.077	345.21	9.49	7.929	8.90	1.70
7	Intermittent viscous debris flow	2.204	476.28	14.53	10.735	8.84	1.50
8	Intermittent viscous debris flow	2.250	425.16	15.49	14.444	7.89	2.00
9	Continuous subviscous debris flow	1.828	99.16	3.075	1.987		
10	Continuous non-viscous debris flow	1.724	43.36	1.178	8.804		
11	Torrential flood	1.400	27.37	0.49	0.439		

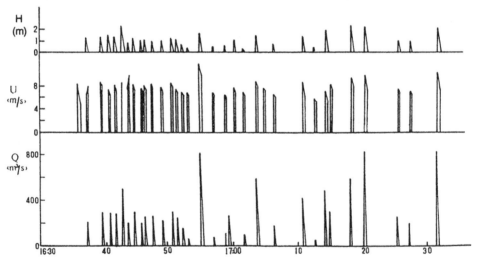

Figure 11.3. A typical process of debris flow waves in the Jiangjia Gully.

tion increases gradually and the flow develops into non-viscous debris flow in 10-20 minutes. Then the flow develops further into subviscous debris flow in a short time. The specific weight of the flowing mixture increases from 1.1 g/cm^3 to 1.9 g/cm^3. Highlight of the process is intermittent viscous debris flow. A series of debris flow waves rush downstream one after another. The density could reach 1.9-2.3 g/cm^3. This process lasts 2-3 hours in general, 80-100 waves pass through in the period. Then the flow transforms into non-viscous debris flow owing to dilution. Table 11.2 presents the dynamic indexes during developing process of a debris flow and Figure 11.3 shows the intermittent process of a viscous debris flow (Kang, 1985b). One can see from the table that following increase in solid concentration or specific weight of the mixture the flow developed from torrential flood into viscous debris flow, the discharge of the flow, the yield stress and rigidity coefficient, and the mean diameter of solid particles increased too. For viscous debris flow as shown in Figure 11.3 the discharge and velocity were intermittent and the velocity of the debris flow wave was as high as 8 m/s, but each wave lasted only about 20-30 seconds. Between the waves the mixture stopped flowing for 1-5 min.

11.3.2 *Paving way process*

Viscous debris flow often develops into a series of waves. When a wave flow through a channel, a layer of water- sediment mixture is stuck on the channel bed. The flow wave becomes smaller and smaller because of the loss of volume of the mixture. It stops at last and forms a way paved with a layer of water-sediment mixture. Then another wave succeeds. This process is termed 'paving way process' by Chinese researchers. Average paving distance of a wave is about 50 m.

11.3.3 *Large gravels concentrate in the surge head*

Each debris flow wave can be divided into three parts: a surge head, a trunk zone and a tail zone. The flow depth decreases from the surge head to the tail zone in

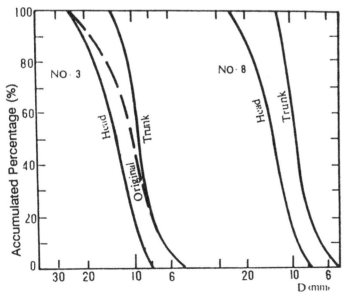

Figure 11.4. Granulometric curves in the surge head and trunk zone of debris flows (Wang & Zhang, 1989).

general. Large gravels and boulders often concentrate in the surge head and form a bulldozer-like front. For example, for debris flow in the Guxiang Gully in Tibet, the surge head of viscous debris flow often consists of boulders up to several meters, while in the trunk zone most of solid particles have a diameter between 0.2 and 0.5 m (Du et al. 1985). Wang & Zhang (1989) conducted a series of experiments of debris flow in a steep flume with a 10 cm-thick layer of 4-25 mm gravels on the bed. Sampling from the surge heads and the truck zones, they got granulometric curves of moving particles in the two zones. Figure 11.4 shows the measured results. The particles in the surge head is apparently greater than that in the trunk zone, in spite of the limitation of the original size distribution.

11.3.4 *Extremely high superelevation at bends and climbing ascend slope*

Some unique natures of viscous debris flow are extremely high superelevation at bends and climbing ascent slope. Table 11.3 gives measured superelevations of viscous debris flows in Japan and United States (Sieyama & Woemoto, 1981; Pierson, 1986). As a comparison, the table also shows the calculated superelevations by using the following equation, which is valid for water flow in rivers,

$$\Delta H = \frac{BU^2}{gR_c} \tag{11.2}$$

where R_c is the radius of the bend, g the gravitational acceleration.

The table shows that the measured superelevation of debris flow was as much as several meters to several tens meters, much greater than the calculated value. Moreover, viscous debris flow can climb a gentle ascent slope, and sometimes can even climb over little hills. When a debris flow wave encounters with a towering obstacle in the flowing course, the front will upraise to a certain height. An empirical formula for the height that a viscous debris flow can ascend or upraise to is given by (Kang, 1985b),

$$H_c = 1.6 \frac{U^2}{2g} \tag{11.3}$$

It is surprising because the maximum height, which a fluid can ascend or upraise

Table 11.3. Superelevation of viscous debris flow.

No.	Location	R_c (m)	U (m/s)	B (m)	$\Delta H_{mea.}$ (m)	$\Delta H_{cal.}$ (m)
1	Yakitakai	442	5	13.5	2.6	0.08
2		212	5	20.0	4.0	0.24
3		74	5	13.0	3.4	0.43
4		94	5	15.0	3.5	0.40
5	Miaokao plateau	365	17	110.0	50.0	8.90
6	St. Helens			70.0	20.0	

to, when all kinetic energy transforms into potential energy without frictional loss, is only $H_o = U^2/2g$.

11.3.5 *Very strong impact force*

Newtonian fluid flowing at a velocity U acts an impact stress, σ, on unit area of a barrier in the course of flow,

$$\sigma = \frac{\gamma_f U^2}{2g} \tag{11.4}$$

where γ_f is the specific weight of the fluid. The formula was also employed in debris flow by some researchers, by substituting the specific weight of the debris

Table 11.4. Impact stress of debris flow.

Time		Velocity U (m/s)	Specific weight γ_m (t/m³)	Impact stress (t/m²)		
				$\sigma_{mea.}$	$\sigma_{cal.\,(11.4)}$	$\sigma_{cal.\,(11.5)}$
July 5	6:24	7.2	2.2	39.0	5.8	15.5
	6:27	6.4	2.2	26.0	4.6	12.2
Aug. 10	16:04	6.9	2.1	28.6	5.1	13.6
	16:22	8.1	2.1	61.2	7.1	18.7
	16:48	5.9	2.1	59.8	3.7	9.9
	17:00	6.0	2.1	23.4	3.9	10.2
	17:04		2.1	104.0		

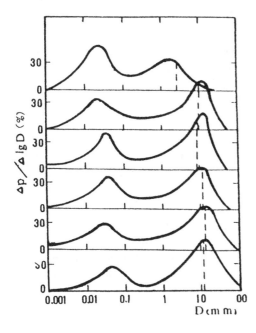

Figure 11.5. Grain size distributions of debris flow in the Liuwan Gully.

flow, γ_m, instead of γ_f. The calculated value from the formula is often much less than measured data. Freshman (1978) reshuffled the formula into:

$$\sigma = 1.33 \frac{\gamma_m U^2}{g} \tag{11.5}$$

Zhang & Yuan (1985) measured the impact stress of debris flow with an electric induction probe at the Jiangjia Gully. Table 11.4 shows a comparison of the measured data with the calculated results from Equations (11.4) and (11.5). The measured impact stress are far greater than the calculated values.

11.3.6 *Bimodal grain size distribution*

Many debris flows have bimodal grain size distribution. Figure 11.5 shows the measured granulometric curves of debris flow in the Liuwan Gully, Northwest China. In the figure $\Delta p / \Delta lgD$ is the frequency density of grains of diameter D. Each curve in the figure has two peak values of frequency density. The first peak is located at $D = 0.01\text{-}0.1$ mm and the second one at $D = 2\text{-}40$ mm.

11.3.7 *Sorting of debris flow deposits*

Generally speaking, the deposit of viscous debris flow consisting of clay, sand and gravels, is a well mixed mixture of various particles. In the Bailong River, the thickness of debris flow deposit is as much as 30-50 m. The profile of the deposit shows huge stones and gravels mixing together with sand and clay (Li & Deng, 1985). But debris flow occasionally brings about inverse grading deposit (upward coarsening deposit), that is quite different from sedimentation in rivers. Such an unique sorting is a character of deposit of surge head of debris flow, in which there is little clay and silt. On the other hand, deposit of non-viscous debris flow is usually positive graded (upward fining). Therefore, deposit of debris flow may be well-mixed, inverse graded, or positive graded.

11.3.8 *Serious degradation and aggradation*

Debris flow often results in great degradation in the upstream and serious aggradation at the mouth of the gully. For instance, in the upstream of the Bomi-Guxiang Gully in Tibet, the gully bed was cut down 140-180 m in the period of 1954-1963, or 16 m annually (Deng, 1985). It was observed that degradation of debris flow gully resulted mainly from retrogressive erosion as debris flow occurred. In a debris flow in the Guxiang Gully in 1964, a 1 m-bed drop moved upstream at a speed of 1 m/min. Sometimes debris flow brings about another type erosion, e.g. large stones in the surge head dig a ditch into the bed and push forward.

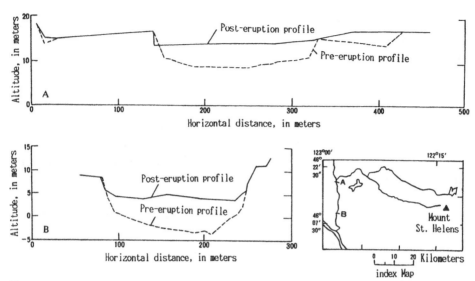

Figure 11.6. Cross-sectional profiles of the Cowlitz River. Preeruption and posteruption of Mount St. Helens (A. 1.0 km below mouth of the Toutle River; B. 11.1 km upstream from confluence with the Colombia River).

As debris flow carries a huge amount of debris to the mouth of the gully, the solid particles will deposit because of sudden reduce of the channel slope. Huge fan-shape stone-sea appears and sometimes the deposit dams rivers. The rate of aggradation is also quite high. It is estimated that more than 1.15×10^8 tones of solid material accumulated and the bed rose 16m in the mouth area of the Guxiang Gully in the period of 1953-1965.

Figure 11.6 shows the aggradation of the Cowlitz River owing to debris flows on May 18 and May 19, 1980 initiated by the eruption of the Mount St. Helens (Janda et al., 1986). The river bed was aggraded 5-8 m after the debris flow.

11.4 MECHANISM OF DEBRIS FLOW

11.4.1 *Matrix and load of debris flow*

Debris flows, except mud flow, are solid-liquid two-phase flow, in which the liquid phase is composed of water, clay and silt while the solid phase consists of sand, gravels and boulders. The liquid phase, or the matrix, of debris flow is essentially a non-Newtonian fluid, usually follows the Bingham constitutive equation (Chapter 5),

$$\tau = \tau_B + \eta \, \frac{du}{dy} \tag{11.6}$$

Because debris flow often has great yield stress, a great amount of fine sediment is carried as neutrally buoyant load. We have discussed in Chapter 2 that the maximum diameter of neutrally buoyant load D_o is given by

$$D_o = k \frac{6\tau_B}{(\gamma_s - \gamma_f)} \tag{11.7}$$

where the constant k is about 1, γ_f here denotes the buoyancy force acting on a particle of unit volume. γ_f equals the specific weight of liquid or clay suspension if a particle suspended in the liquid or clay suspension. In debris flow, however, there are solid particles with various diameters. It is revealed that only small particles make contribution to the buoyancy force of a suspended particles (see Chapter 6). As a rough approximation γ_f can be taken as the specific weight of the mixture excluding grains larger than $D_o/50$ (Wang, 1987).

In non-viscous debris flow, turbulent diffusion is still intense and clay concentration is low. The solid particles of sand size move as suspended load rather than neutrally buoyant load, while gravels and boulders move as bed load.

In viscous debris flow, yield stress and viscosity of the matrix is high. Turbulent eddies are suppressed by the yield stress and high viscosity and the flow becomes laminar. The maximum diameter of neutrally buoyant load D_o can be as large as several millimeter. Fine and coarse sand is supported by the strength of the matrix and becomes neutrally buoyant load. The flow carries more gravels and boulders owing to increase in competency of the flow. They move as bed load, or laminated load.

Wang et al. (1989) studied the structure of clay mud and found from experiments that clay mud exhibits not only yield shear stress, but also yield normal stress. With a specially designed instrument, they found that when a pushing force exerted on clay mud, a probe plate in the mud at a certain distance measured a normal stress. Figure 11.7a shows the set-up for measuring the yield normal stress σ_m. The probe plate was hung on an electronic balance. As weights were added into the cylinder which was over the plate, reading on the balance increased because mud between the plate and the cylinder transmitted the normal stress exerted by the cylinder to mud. The value of the normal stress varied with the magnitude of the exerted stress, the distance of the probe plate from the bottom of the cylinder, and mineralogy and concentration of the clay mud. If the normal stress acting on the mud was too large, the structure of the clay mud would be destructed and the cylinder would settle down. Figure 11.7b shows the maximum normal stress σ_m received by the probe plate at a distance of 1 cm for two kinds of clays at different concentrations. It can be seen that the value of σ_m is much larger than yield shear stress τ_B. The implication of the phenomenon will be discussed latter.

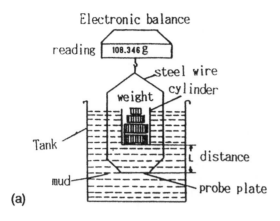

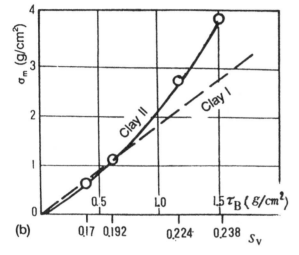

Figure 11.7. Set-up for measurement of yield normal stress (a); and yield normal stress as a function of concentration and yield shear stress (b).

11.4.2 *Movement of coarse grains*

In viscous and subviscous debris flows, there is little or no suspended load, particles of a diameter larger than D_o move as bed load (laminated load). The transport rate of the bed load in debris flows follows the rule of general bed load motion. According to Qian (1980), most of the formulas for transport rate of bed load can be transformed into a correlationship between the bed load transport intensity Φ, and the flow intensity, Θ. In debris flow, Φ and Θ can be expressed as

$$\Phi = U_b H S_{vb} \left(\frac{\gamma_f}{\gamma_s - \gamma_f} \right)^{1/2} \left(\frac{1}{gD_b^3} \right)^{1/2} \tag{11.8}$$

$$\Theta = \frac{\gamma_f R J}{\gamma_s - \gamma_f} \tag{11.9}$$

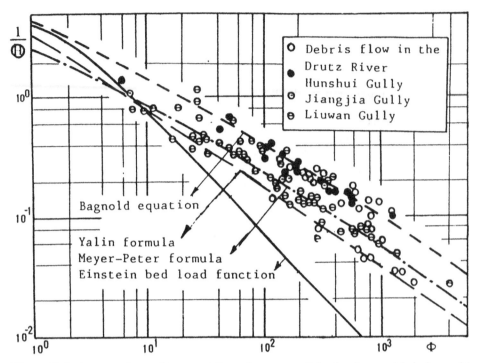

Figure 11.8. A comparison of transportation of coarse particles in viscous debris flow with bed load formulas.

where U_b is the average velocity of bed load particles, S_{vb} the average volume concentration of bed load; R hydraulic radius; H depth; J energy slope; D_b mean diameter of the particles coarser than D_o.

Four typical formulas of rate of bed load transport and measured data of viscous debris flows are plotted in Figure 11.8 (Qian & Wang, 1984). The measured points fall in between Meyer-Peter formula, Bagnold equation, and Yalin formula. It indicates the movement of coarse particles in debris flows still obeys the same rule as bed load transport in sediment-laden flow. Einstein bed load function differs from the other three formulas as $\Phi > 10$ and deviates from measured points because Einstein bed load function was established on the basis of a continuous interchange and balance between moving particles and particles resting on the bed. Coarse grains in debris flow move essentially as laminated load, no exchange occurs between moving particles and bed particles. Einstein bed load function, therefore, fails to be applied in the case of debris flow.

11.4.3 *Mechanism of bimodal grain size distribution*

As mentioned in Section 11.3, debris flow, especially viscous debris flow, has bimodal grain size distribution. Characteristics of the grain size distribution are,

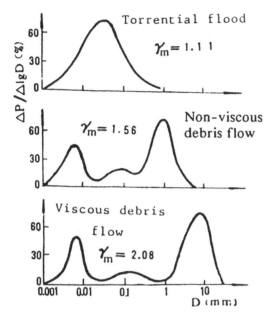

Figure 11.9. Patterns of grain size distribution in different stages in developing process of debris flow in Jiangjia Gully.

of course, closely related to the supply conditions of sediment in the watershed. On the other hand it is also affected by the selectivity of moving pattern to solid grains in the course of movement.

Figure 11.9 shows the grain size distributions in a developing process from torrential flood, non-viscous debris flow to viscous debris flow occurred in the Jiangjia Gully (Kang, 1985b). Suspended load dominated in the torrential flood and the size distribution had only one peak of frequency density. In non-viscous debris flow, laminated load and neutrally buoyant load coexisted with suspended load. Two new peaks emerged in the distribution curve, adding to the lowered peak corresponding to suspended load. With sediment concentration increasing further, the flow turned into viscous debris flow. Laminated load dominated and corresponded to the higher peak in the size distribution curve. Suspended load motion vanished. Fine sediment moved as neutrally buoyant load corresponding to the lower peak in the neighborhood of $D = 0.01$ mm. Analysis of data collected from the debris flow in the Liuwan Gully indicated about 50-60% of the solid grains moved as laminated load (Qian & Wang, 1984). The diameters marked by dotted lines in Figure 11.5 are representative diameters of laminated load D_b which coincide with peak at about $D = 10$ mm. This proves that correlation exists between grain size distribution and sediment movement patterns.

11.4.4 *Paving way process*

Viscous debris flow has yield stress, and flows down an inclined channel only if its depth exceeds a critical value H_o. When a viscous debris flow wave passes through an open channel with a slope J, the depth H in the surge head is large enough but it is smaller and smaller from the head to the tail. A tail section with depth less than H_o stops moving under action of the yield stress. The wave, therefore, becomes shorter and shorter and rests on the channel bed at last.

Denote λ the length of the debris flow wave, H the average depth and U the average velocity. $B\rho_m gHJ\lambda$ and $\tau_o B\lambda$ are driving force and resistance acting on the wave, respectively, where B is the width of the channel. The law of momentum conservation yields

$$\frac{\partial MU}{\partial t} = \rho_m gHJB\lambda - \tau_o B\lambda \qquad (11.10)$$

where M is the total mass of the debris flow wave. In paving way process, a layer of debris mixture of thickness H_o is stuck on the bed when a wave flows through a channel. The wave flows downstream a distance U in unit time. The momentum of the wave decreases by $\rho_m H_o U^2 B$ in unit time. Equation (11.10) can be rewritten as

$$-\rho_m H_o U^2 = \rho_m gHJ\lambda - \tau_o \lambda \qquad (11.11)$$

Take $\tau_o = \tau_B$ and $H = H_o$ for simplicity, which is reasonable for viscous debris flow wave because the yield stress is high and the flow is laminar. H_o can be solved from Equation (11.11) (Wang et al. 1989),

$$H_o = \frac{\tau_B}{\rho_m gJ + \dfrac{\rho_m U^2}{\lambda}} \qquad (11.12)$$

Equation (11.12) interprets that the thickness of sticking debris mixture on the channel bed not only depends on the yield stress, but also varies with J and U^2/λ. H_o reaches the maximum value H_{om} when $U^2/\lambda = 0$,

$$H_{om} = \tau_B/(\rho_m gJ) \qquad (11.13)$$

H_{om} equals the maximum depth that a Bingham fluid can rest on an inclined plane with slope J.

If the total volume of a wave is V_o, the length of the way it can pave is L

$$L = \rho_m V_o \frac{gJ + \dfrac{U^2}{\lambda}}{\tau_B B} \qquad (11.14)$$

Examination with measured data of debris flow and experiments fully proved above conclusions. The higher the velocity U and the smaller the yield stress, the smaller the depth H_o and the longer the paving distance L. It is also concluded that as long as the matrix of debris flow exhibits yield stress, the first few debris flow waves have to pave the way and then the following waves can flow through the channel.

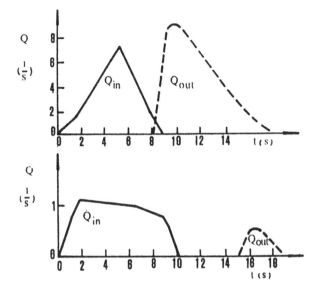

Figure 11.10. Growing and reducing of clay mud waves ($J = 0.07$, $S_v = 0.18$).

Because the depth of sticking debris mixture H_o on the channel bed varies with velocity, a debris flow wave can grow up if its velocity is high, and it can also reduce if its velocity is low when the wave passes through a channel paved with debris mixture by previous waves.

Experiments were conducted in a flume to study the paving way process. Figure 11.10 shows variation of discharge of two waves flowing in the flume (Wang et al., 1989). In which Q_{in} and Q_{out} are discharges at the entrance and at the downstream end of the flume, respectively. The experiments were carried out with clay mud. Waves with different discharges flowed into the flume one by one, the 'paving way process' took place first, nothing flowed out of the flume at this time. As the whole flume was stuck by a layer of mud, following waves passed through the flume. A mud wave with a maximum discharge 7.2 l/s entered into the 8.7 m-long flume, it became a larger wave with a maximum discharge 9.2 l/s, while a small wave with a maximum discharge 1.15 l/s became a still smaller one ($Q_{out} = 0.5$ l/s) when it flowed out of the flume.

Debris flow wave can grow up or reduce depending on the thickness of debris mixture layer sticking on the channel bed previously. This brings about great difficulty in predicting the debris flow discharge. Figure 11.11 gives the variation of discharge of a debris flow wave along the flowing course in the Little Almakinca River. The discharge was less than 20 m³/s and increases to 150 m³/s when it traveled 4 km. It varied from 150 m³/s to about 60 m³/s when it traveled 0.4 km farther, and then it increased to about 230 m³/s in another 1.6 km journey.

Let H_{o1} be the thickness of debris mixture layer sticking on the channel bed previously, L be the length of the channel. A debris flow wave of total volume V_o flows into the channel at velocity U. The thickness of debris mixture layer sticking

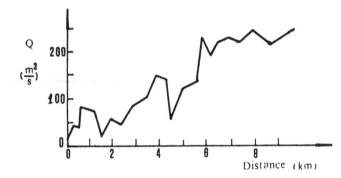

Figure 11.11. Variation of discharge of a debris flow along the Little Almakinca River.

on the channel bed will become H_{o2} after the wave passed through. The total volume of the wave will change to V, which is given by:

$$V = V_o + BL \left[H_{o1} - \frac{\tau_B}{\rho_m gJ + \rho_m \dfrac{U^2}{\lambda}} \right] \tag{11.15}$$

If H_{o1} and velocity U are large, the second term on the right hand side of Equation (11.15) is positive, the debris flow wave will expand. If H_{o1} and velocity of the wave are small, the second term is negative and the wave will reduce. Especially, if $H_{o1} = 0$, paving way process will occur and the equation reduces to formula (11.14) in this case.

11.4.5 *Structure of debris flow wave*

Pierson (1986) studied dynamic and physical properties of channelized debris flows at Mount St. Helens in 1981-1983. He collected data of velocity, flow depth, and composition of solid materials by using motion picture photography, time lapse photography, acoustic rangefinding, timing drift, and hand sampling. He found that debris flow waves ranged in magnitude from 1 to 50 m³/s. Each flow wave had a very steep surge head, containing the densest slurry and reaching the highest stage of flow. The head impeded the flow of more fluid slurry behind the front, and a progressively more dilute tail that accounts for the recessional limb of the 'slurry flood wave'. Front velocity was less than thalweg surface velocity in general. For instance, the front of a debris flow occurring on Sept. 3, 1982 moved at a speed of 2.3 m/s while the thalweg surface velocity was as much as 3.8 m/s and the average velocity in trunk zone was estimated as 3.4 m/s.

Figure 11.12 shows a typical debris flow structure. The steep, lobate surge head was composed predominantly of the coarsest particles available for transport. Such head was typically an openwork pile or ridge of boulders being bulldozed along by the flow. When matrix slurry did not fill all the interstices, the head was not liquefied and internal friction was much higher than in the rest of the debris

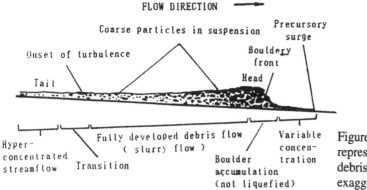

FLOW DIRECTION ⟶

Coarse particles in suspension   Precursory surge

Onset of turbulence   Bouldery front

Tail   Head

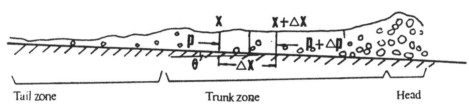

Hyper-concentrated streamflow   Fully developed debris flow (slurry flow)   Transition   Boulder accumulation (not liquefied)   Variable concentration

Figure 11.12. Schematic representation of a typical debris flow (vertically exaggerated).

X   X+ΔX

P   P+ΔP

θ   ΔX

Tail zone   Trunk zone   Head

Figure 11.13. A model of debris flow wave.

flow. Motion picture photography of boulder movement on the surface of surge head showed that boulders on the surface were moving 1.8 times faster than the head itself. Surface particles moved to the front and tumbled down the steep leading edge. Cobbles and small boulders were overridden and reincorporated into the flow; the large boulders were simply pushed ahead.

Wang & Qian (1989) presented a model of the structure of debris flow. According to the foregoing discussion on properties of matrix of viscous debris flow, the matrix of high concentration of clay can deliver a pushing force to a certain distance (Wang et al., 1989). On the other hand, solid particles (gravels and boulders) collide with each other continuously. Collision of grains in front and behind results in momentum exchange and a pushing force. These forces make the movement of one part of the wave be affected by other parts of the wave.

Consider a viscous debris flow wave flowing downstream a gully with a rectangular cross-section. Take a volume element bounded by $x = x$ and $x = x + \Delta x$ as shown in Figure 11.13. The energy balance equation gives

$$\rho_m BHU\Delta x \frac{dU}{dt} = \rho_m gHBU\Delta x \sin\theta - \tau_o(B + 2H)U\Delta x - UB\Delta(PH)$$

(11.16)

Where P is the resultant pushing force owing to the strength of matrix and

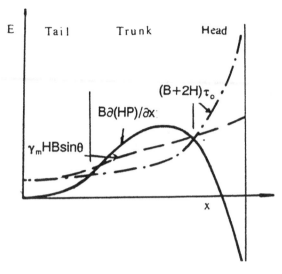

Figure 11.14. Energy structure of a debris flow wave.

collisions of solid grains. The left hand side of Equation 11.16 is inertial energy and the first term on the right hand side is the energy supply from gravitation in unit time, the second term is the energy consumed locally and the third term is the energy delivered to the surge head. Divided by $BU\Delta x$, the equation can be reshuffled into

$$\rho_m BH \frac{dU}{dt} = \rho_m gBH \sin\theta - \tau_o(B + 2H) - B\left(H\frac{\partial P}{\partial x} + P\frac{\partial H}{\partial x}\right) \qquad (11.17)$$

In steady flow, the left hand side of Equation (11.17) is zero, and the distributions of the rest three terms along the wave length are shown qualitatively in Figure 11.14.

The surge head of a debris flow wave moves like a bulldozer. Gravels, pebbles and boulders in the head strike against the channel bed when they fall from the top of the head, which results in great energy consumption. The energy consumption during movement of the head is much more than the energy supply. To maintain a steady movement, the trunk zone delivers energy to the surge head. In fact the debris flow head is pushed forward by the trunk zone.

It was observed from a debris flow experiment that liquid and gravels in the trunk zone moved much faster than the surge head. The gravels caught up with and roll over the head and then stopped over on the bed. Liquid in the trunk zone flowed through the surge head and delivered its kinetic energy to gravels and then fell on the bed and scoured bed gravels. The surge head rolled forward. Collision of gravels in the head made noise and consumed a lot of energy. The liquid flow functioned like a energy conveyer belt and transferred energy to the head continuously. If the supply of liquid was cutted just after a debris flow had been initiated and a surge head formed, the head would reduce gradually and gravels in the head stop moving (Wang & Zhang, 1990). This result supported the theory of the longitudinal structure of debris flow wave.

As a debris flow wave runs into an impediment abruptly, such as an obstacle, ascent slope or a sharp bend, the resistance force acting on the head increases abruptly. In the trunk zone, however, the terms of energy supply and local energy consumption in Equation (11.17) do not change at this time, the third term in Equation (11.17) is bound to increase because du/dt is negative. This means that a part of inertial energy of the trunk zone is transmitted to the head when speed of the wave decreases sharply.

The property of energy transmission endows the debris flow wave some abnormal behaviors. For example, the head of debris flow wave, pushed by the trunk zone, can climb a gentle ascent slope. When a debris flow wave encounters with a towering obstacle in flowing course, the surge head can upraise to a height larger than $H_o = U^2 /2g$. Superelevation of viscous debris flow at bends is distinctly higher than that of water flow. It is also attributed to the structure of debris flow wave. The pushing force of the trunk zone acting on the head highly depends on the strength of the matrix and collisions of large particles, and results in greater impact force on barrier than the values given by Equations (11.4) or (11.5).

11.4.6 *Mechanism of large particles concentrating in the surge head*

Takahashi gave an explanation of the phenomenon of larger particles concentrating in the surge head: The dispersive force resulting from collisions of particles is proportional to square of diameter of particle, so that large particles can rise gradually to the surface of the flow where the flow velocity is greater than that bellow. Therefore large particles move faster than small ones and concentrate in the head at last (Takahashi, 1980).

Wang & Zhang (1990) designed a special experiment to investigate the mechanism by using tracer particles. They observed that large gravels moved faster than small particles but did not rise to the upper flow. The particle's size in the upper flow was not greater than that in the lower flow. Large gravels rolled on the bed at a stable velocity and their movement was little affected by collisions with small particles, while small particles moved in saltation, sometimes jumped into upper zone and moved at a high speed and sometimes fell in lower zone and moved at a low speed. Once a small particle collided with a large gravel, it might slow down sharply or possibly stop moving. Gravels were acted by a tractive force from the liquid phase. The tractive force could be roughly estimated with the formula:

$$F = 3\pi\eta_e Du_r, \text{ for Re}_r \leq 1$$
$$F = k\rho_f D^2 u_r^2, \text{ for Re}_r > 1000 \tag{11.18}$$

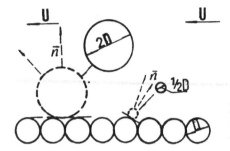

Figure 11.15. A sketch resistances of larger and smaller particles owing to collision with bed particles.

where u_r is the relative velocity between gravel and ambient liquid, k a constant, ρ_f the density of the liquid, η_e the effective viscosity of the liquid which can be written as:

$$\eta_e = \eta + \frac{\tau_B}{\left(\dfrac{du}{dy}\right)} \tag{11.19}$$

for Bingham fluid. The Reynolds number Re_r is defined by:

$$\mathrm{Re}_r = \frac{\rho_f D u_r}{\eta_e} \tag{11.20}$$

Equation (11.18) indicates that the tractive force acting on gravels is proportional to gravels' diameter or to square of the diameter. The larger the particle, the greater the tractive force. On the other hand, the resistance force acting on large particle is not larger than that on small particle. The resistance force on particles in debris flow mainly comes from collisions with bed gravels or with lower moving gravels. Collision of larger gravel with bed gravels does not change the gravel's momentum in the flow direction a lot because the contact plane does not face backward, but a smaller particle may totally lose its momentum component in the flow direction at one collision with bed gravel as shown in Figure 11.15. Large gravels are acted by larger tractive force and smaller resistance force, therefore, they move faster than small ones and concentrate in the surge head.

11.4.7 *Initiation of debris flow*

Using the concept of dispersive force, Bagnold (1956) attempted to explain various sediment transport processes including debris flow. He established the following formula to illustrate the condition of initiation of debris flow,

$$HS_v \tan \alpha = \frac{\rho_f}{\rho_s - \rho_f} H \tan \theta + HS_v \tan \theta \tag{11.21}$$

where $\tan \theta = J$ is the slope of the bed, α dynamic angle of internal friction. The left hand side of Equation (11.21) is the internal resistance shear stress and the right hand side is tractive shear stress. In the case that the left hand side of Equation (11.21) is greater than the right hand side, a layer of grains on the bed will remain stationary. However, when the left hand side is less than or equal to the right hand side, a simultaneous shearing of the grain layer will take place and the grain layer begins to move. This is essentially the condition for the initiation of debris flow, triggered by overland flow on the grain bed.

We can solve $\tan \theta$ from Equation (11.21) which refers to the critical slope for initiation of debris flow

$$\tan \theta = \frac{S_v \tan \alpha}{S_v + \dfrac{\rho_f}{\rho_s - \rho_f}} \tag{11.22}$$

Bagnold (1956) calculated θ values for a number of possible cases: (1) For slowly flowing quartz grains with $S_v = 0.6$ and $\tan \alpha = 0.63$, Equation (11.22) gives $\tan \theta = 0.33$ or $\theta = 18\text{-}19°$. (2) For rapidly moving grains, supposedly in the grain-inertia region, the bed tends to disperse slightly to its fluidization limit with $S_v = 0.53$ and $\tan \alpha = 0.32$ and then Equation (11.22) yields $\tan \theta = 0.15$ and $\theta = 8\text{-}9°$.

Takahashi (1980) further developed Bagnold's theory. He analysed the equilibrium condition between the tractive shear stress and the resistance shear stress over the whole layer of debris deposits when torrential flood flows on the surface. He obtained the critical slope for initiation of debris flow

$$\tan \theta = \frac{S_{v*}(\rho_s - \rho_f)}{S_{v*}(\rho_s - \rho_f) + \rho_f(1 + H_o/D)} \tan \alpha \tag{11.23}$$

in which H_o is the depth of the overland flow on the debris layer, S_{v*} is the volumetric concentration of solid grains of the debris layer. Taking $S_{v*} = 0.7$, $\rho_s/\rho_f = 2.6$, $H_o/D = 1.4$ and $\tan \alpha = 0.8$, Takahashi calculated $\theta = 14.3°$.

Wang & Zhang (1990) studied the initiation of debris flow experimentally, by allowing water or clay mud, with different clay concentrations and at different flow rates, to flow over a flume bed piled with gravels ($D = 4\text{-}25$ mm), and observing and measuring movement of gravels, flow of mud and interaction of the two phases. The experiment was conducted in an 8.7 m-long, 10 cm-wide and 20 cm-deep plexiglass flume. The bed slope could be adjusted in the range of $J = 0\text{-}20\%$. The liquid phase was water or mud with different concentrations of clay. The rheological behavior of the clay suspension approximately followed the Bingham model.

If the flow rate was high enough the liquid flow was turbulent and it could trigger a mass of gravels to enter into motion. Many gravels concentrated in front of the flow and formed a steep, high surge head. The speed of the head was much lower than the velocities of the liquid phase and gravels in the trunk zone. If the flow rate was low and the slope of the channel was gentle or the clay concentration was very high, the flow became laminar and very few gravels were picked up from the bed. The front of the flow was low, flat and free of gravels in this case. The speed of the front was higher than that of the liquid flowing behind. The former described above is referred to debris flow and the latter is attributed to ordinary sediment-laden flow. Initiation of debris flow depends mainly on flow rate of liquid in unit width, q, and slope of the channel bed. Figure 11.16 shows critical conditions for initiation of debris flow, in which data from Takahashi (1978) are also plotted for comparison. It can be seen that the curves to divide the zone of initiation of debris flow and the zone of non-initiation in q-J coordinates plane for different liquid phases are hyperbolas, which suggests a criterion of initiation of debris flow, $K = \gamma_f qJ$. The critical condition for initiaton of debris flow is given by

$$\gamma_f qJ > K_c \tag{11.24}$$

where K_c depends on the liquid phase. $\gamma_f qJ$ can be interpreted as energy of the

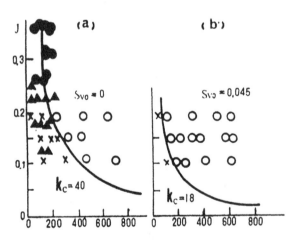

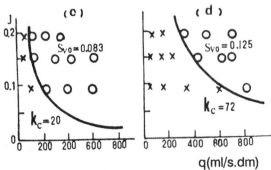

Figure 11.16. Critical conditions for initiation of debris flow:
o debris flow by Wang & Zhang (1989); • debris flow by Takahashi (1978); x no debris flow by Wang & Zhang (1990); ▲ no debris flow by Takahashi (1990).

q(ml/s.dm)

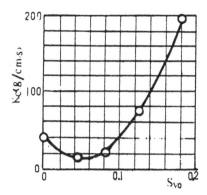

Figure 11.17. Criterion K_c as a function of clay concentration.

liquid phase supplying to the flow in unit time and K_c is the minimum energy for triggering debris flow. As shown in Figure 11.16, K_c varies with clay concentration S_{vo}. Figure 11.17 shows K_c as a function of S_{vo}. Gravels in the flow moved as bed load, they collided with each other consecutively during their movement. The collisions resulted in a dispersive force which supported their effective weight, and at the same time resulted in a large resistance force. The liquid phase acted a tractive force on the gravels to balance the resistance force, which is given by Equation (11.18). Because η_e increases quickly with increasing clay concentration, increase in S_{vo} enhances the tractive force on gravels and consequently reduces K_c, so that K_c reduces with increasing S_{vo} if S_{vo} is less than 0.06. As S_{vo} is larger than 0.06, however, K_c increases with increasing S_{vo} fast that means that the thicker the mud is, the more difficult the initiation of debris flow will be. This can be interpreted from the fact that if the concentration is high, the flow remains in laminar state. Energy diffusion in laminar flow is much weaker than in turbulent flow. Only the liquid near the bed and contacting with gravels delivers kinetic energy to gravels, and the upper part of the flow contributes little energy to bed load movement and flows at a high velocity. Consequently a higher total energy of the flow is needed for triggering debris flow.

The smallest K_c falls in the range of S_{vo} = 0.04-0.08 which indicates that a flow of liquid with clay concentration in this range most likely initiates debris flow. It fact, most of debris flows occurring in nature have liquid phase of clay concentration in the range of S_{vo} = 0.04-0.09 which conforms the experimental results. The flow of mud with clay concentration in the range is turbulent, it scours the gravel bed into a considerable depth. The liquid flow in the lower zone consumes a lot of energy on carrying gravels while the turbulent diffusion of momentum transports energy from the upper zone to the lower zone. Therefore the flow can initiate debris flow at a lower flow rate and smaller bed slope.

It must be pointed out that the high concentration clay mud has high competence to carry solid grains so that high clay concentration of matrix enhances the stability of debris flow and counteracts deposition of solid grains. Therefore, as long as a debris flow is initiated the more clay and silt join the flowing liquid, the more stable the debris flow.

11.4.8 *Velocity and velocity profile*

The Chezy's formula in hydraulics is often used in estimating the average velocity of debris flow.

$$U = C\sqrt{RJ} \tag{11.24}$$

$$C = \frac{R^y}{n} \tag{11.26}$$

where C is the Chezy coefficient, n Manning roughness of the debris flow gully.

Kang (1985a) analysed the field data of about 1000 debris flow waves in the Jiangjia Gully measured in the period of 1966-1975 and got the following empirical formulas of average velocity

$$U = \frac{1}{n} H^{2/3} \sqrt{J} \tag{11.27}$$

$$\frac{1}{n} = 28.5 \, H^{-0.34} \tag{11.28}$$

where H is the depth of the debris flow. The roughness n increases with increasing H because the larger the depth H, the higher the specific weight of the debris mixture and the coarser the grains carried in the flow.

Similar relationship between roughness and depth was also found in other debris flow gullies. For example, for debris flow in the Huoshao Gully, Gansu province,

$$\frac{1}{n} = 23.4 \, H^{-0.397} \tag{11.29}$$

Table 11.5. Manning roughness of debris flow gullies.

Characteristics of debris flows	Channel bed	Roughness n
Solid grains mix with mud, diameter of grains ranges from 1 mm to several meters. D_{50} is about 300-500 mm, particles of 2-5 m are about 20% of the total	Channel bed is very rough, there are many huge stones, over 100-1000 m³. Slope of the bed is in the range of 0.1-0.15	Average value of n is 0.27, and when $H > 2$ m, n increases to 0.445
$D_{50} = 200$-300 mm, very few particles are larger than 2 m	Channel bed is full of bumps, its slope is about 0.07-0.1	For $H < 1.5$ m, $n = 0.05$-0.33, and for $H > 1.5$ m, $n = 0.05$-0.1
$D_{50} = 100$ mm, different particles mix with liquid uniformly, the flow is turbulent	Gravels on the bed are quite uniform and about 100 mm. The slope is about 0.055-0.070	For $0.1 < H < 0.5$ m, $n = 0.043$, for $0.5 < H < 2$ m, $n = 0.077$, and for $2 < H < 4$ m, $n = 0.10$
Ditto	After the paving process, the bed becomes smoother than before	For $0.1 < H < 0.5$ m, $n = 0.022$, for $0.5 < H < 2$ m, $n = 0.038$, for $2 < H < 4$ m, $n = 0.05$

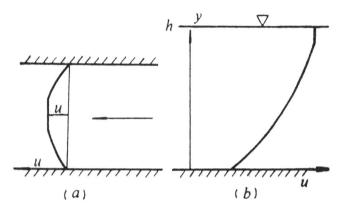

Figure 11.18. Surface velocity profile (a) and vertical velocity profile (b).

and in the Niwan Gully, Gansu Province,

$$\frac{1}{n} = 18\,H^{-0.403} \tag{11.30}$$

Zhang et al. (1985) tabulated the roughnesses of debris flows in different cases, as shown in Table 11.5.

The velocity profile of debris flow is very difficult to measure. What one can do now is using video tape camera and camera to take photos and observe the surface flow and analyze the surface velocity distribution in a transverse section and the vertical velocity profile of gravels in the front of the surge head. Figure 11.18 shows typical surface profile and vertical profile in the Jianjia Gully (Kang, 1985a).

Pierson (1986) reported that the speed of surge head was much lower than liquid velocity or gravel's velocity in trunk zone. The same phenomenon was found in debris flow experiment (Wang & Zhang, 1990). Table 11.6 presents propagation speed of debris flow head, U_b, and relative parameters, in which u_s is the surface velocity of liquid in the trunk zone, H and H_b are average flow depth in the trunk zone and height of the surge head, respectively; S_b and D_{b50} are concentration (weight/total volume) and median diameter of gravels in the head; S and D_{50} are concentration and median diameter of gravels in the trunk zone. The propagation speed of debris flow is only one quarter to one half of the surface velocity of the liquid phase. The speed of the head U_b increases with increasing q and decreases with increasing H_b. This is because the larger the q, the more energy is transported from the trunk zone to the head; and the larger the H_b, the more energy is consumed by the head in its movement. Figure 11.19 shows the head speed U_b as a function of q/H_b. It is obvious that U_b varies linearly with q/H_b, and the larger the slope J and the concentration S_{vo}, the steeper the lines.

Takahashi and Wang measured the vertical velocity distribution of grains and the liquid phase in flume experiments of grain flow. The results may be applied for water-rock debris flow (see details in Chapter 7).

Table 11.6. Velocity of debris flow and relative parameters.

J	S_{v0}	q (ml/s.cm)	U_s (m/s)	U_b (m/s)	S_b (kg/m³)	S (kg/m³)	H_b (cm)	H (cm)	D_{b50} (mm)	D_{50} (mm)	q/H_b (ml/s.cm)
0.103	0.045	140	0.85	0.21	540	317	4.0				35
		250	1.21	0.49	533	432	5.0	4.0			50
		420	1.77	0.66	807	477	6.3	5.0			67
		600	1.83	0.68	814	573	7.1	5.8			85
0.096	0.079	190	1.46	0.56	528	271	3.8	2.5			50
		410	1.83	0.85	402	321	4.8				
		620	2.02	0.99	949	331	5.5				
		730	2.21	1.21			6.0				
0.152	0.045	200	1.26	0.41	1119	785	4.8	3.5	12.7	11.4	42
		350	1.32	0.66	1299	927	5.5	3.8	13.0	10.4	64
		630	2.21	0.85	1154	827	6.5	4.8	11.2	9.2	97
		570	2.27	0.79	1241	944	6.0		11.0	9.7	95
0.152	0.079	85	0.79	0.32	805	565	4.0				22
		180	0.97	0.33	847	548	5.0	3.0	14.1	10.6	36
		285	1.31	0.62	1338	866	5.2		12.0	10.5	55
		440	1.77	0.72	872	615	5.6	4.0	13.4	9.7	79
		590	2.50	1.20	1406		6.5	5.0	13.0	9.3	91
0.154	0.125	150	1.72	0.86	873	545	4.0	2.5			37
		400	1.82	1.09	769	635	5.1	3.5			80
		600	2.51	1.28	857	653	6.6	5.0			91
		650	2.54	1.64	765	514	6.6	5.0	11.8	10.0	98
0.194	0	430		0.69	1890		5.5	3.0			78
		620		1.15			6.2	5.0			100
0.191	0.045	510	3.21	1.30	2000		7.2	4.0			71
		300		0.81	2000		5.8	2.5			52
0.194	0.086	180		0.90	1700		5.0	2.5			36
		320		1.00	2010		7.0	3.0			45
0.191	0.132	320	1.74	0.99	1090	851	5.5	3.0			58
		500	2.26	1.29	1218	792	6.8	3.5			74
0.191	0.125	715	2.35	1.32	1216	625	7.6	5.0	12.2	10.3	94

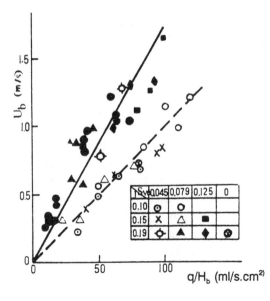

Figure 11.19. Speed of surge head versus q/H_b.

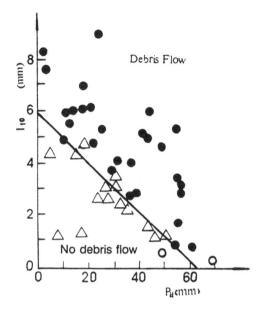

Figure 11.20. Critical rainfall for debris flow.

11.5 PREDICTION, PREVENTION AND CONTROL OF DEBRIS FLOW

11.5.1 *Prediction of debris flow*

Debris flow often brings about serious disaster and great loss of people's properties. Prediction of debris flow is of great importance and remains in developing stage. An important routine of prediction of rainstorm debris flow is to relate rainfall intensity in 10 minutes and the antecedent rainfall to initiation of debris flow. Chen (1985) analysed data of nearly one hundred rainfalls and tens of debris flows in the watershed of the Jiangjia Gully. The main results he obtained are shown in Figure 11.20, where I_{10} is 10 minutes rainfall intensity, or the maximum precipitation in 10 minutes, P_a is a factor of accumulated precipitation in recent 20 days and is given by the following formula:

$$P_a = P_0 + 0.8\,P_1 + 0.8^2 P_2 + 0.8^3 P_3 + ... + 0.8^{20} P_{20} \qquad (11.31)$$

in which P_0 is the precipitation just before the most intensive 10-min. rainfall, P_i is the precipitation on the day i-days before that day. Figure 11.20 shows that debris flow occurs if:

$$10.33\,I_{10} + P_a > 62\ \text{mm} \qquad (11.32)$$

The analysis also indicates that if P_a is smaller than 30 mm and I_{10} is larger than 4 mm, there occurs gusty and short-time debris flow, while if P_a is larger than 60 mm and I_{10} is less than 2 mm, non-viscous debris flow occurs. By combining the results and rainstorm forecasting, debris flow in that area can be predicted. The same method and the criterion I_{10} were employed for prediction of debris flow in other debris flow gullies (Liu & Cheng, 1983).

11.5.2 *Prevention from debris flow disasters*

Many detecting and warning systems have been developed for preventing railways, highways, bridges, factories and mines from debris flow disasters. These detecting and warning systems work based on different principles. For instance, vibration detector receives vibration induced by debris flow and transmits warning signal to the protected objects. Debris flow level detector can send warning signal to the protected objects when a debris flow is over a given stage. These warning systems are working in many debris flow areas, they have successfully sent warning signals to railways, bridges and towns in time and saved people's lives and properties.

11.5.3 *Control engineering*

Debris flow-control engineering has a long history in China. People dwelling in debris flow area built dams and dredged flood-ways in order to avoid or reduce debris flow disasters several hundreds years ago. Engineering measures for controlling debris flow can be classified into following several kinds:

(a) *Diversion works in the upstream of debris flow gully*. These diversion works divert flood from the upperstream of debris flow gully, and reduce discharge of flood and kinetic energy of the fluid so that debris flow is prevented or reduced in scale.

(b) *Dam and dam-train*. Dam and dam-train on debris flow gully were built to trap debris and effectively check debris flow. The Daqiao Creek lying on the right side of the Xiaojiang River is a very active debris flow gully. Five detention dams were built in 7 km along the gully. These dams reduced solid material transported into Xiaojiang River considerably and they were enhanced step by step whenever the reservoirs formed by the dams were silted up. Nowadays lattice dam is widely used because it can trap large boulders in debris flow and reduce harmfulness, and on the other hand it has much longer life span because it is silted up at a very low speed.

(c) *Pilot trench and flume*. Debris flow from the Jiangjia Gully dammed the Xiaojiang River many times. To control debris flow a 2.5 km-long diversion channel was built to divert the debris flow to right side of the fan area. The right side of the diversion channel is by a mountain which formed a wall of the channel while the left side of the channel was made of mud-rock which was transported by previous debris flow. The diversion channel has effectively guided debris flows to the deposited area and has prevented the Xiaojiang River from blocking for twenty-two years.

To built flume across railway or highway is an effective measure and is widely used in China. Only in Gansu Province there are 25 flumes which guide debris flow across over the highways or railways and consequently protect the highways and railways from damage of debris flow.

(d) *Debris basin.* Another debris flow control measure is to establish debris silt basin. By guiding debris flow into a given area and trapping debris farm land, hydraulic works and dwelling area are protected from disaster of debris flow.

(e) *Stabilizing slope.* Reforestation is an ecologic measure for controlling debris flow thoroughly. In Nanpin county, Sichuan province, for example, people stop ploughing and growing crops in slope-land for reforestation, and planting grass. This effectively reduced debris flow disasters. Building soil-retaining structure at particular places is necessary for preventing landslide and debris flow. It is also benefit to the formation of a protective cover of vegetation.

REFERENCES

Bagnold, R.A. 1954. Experiments on gravity-free dispersion of large spheres in a Newtonian fluid under shear. *Proc. Royal Soc. London, Ser. A*, Vol. 225.

Bagnold, R.A. 1956. Flow of cohesionless grains in fluid. *Philos. Trans., Royal Soc. of London, Ser. B*, 249: 235-297.

Chen, J. 1985. *A Preliminary analysis of the relation between debris flow and rainstorm at the Jiangjia Gully of Dongchuan in Yunnan* (in Chinese). Memoirs of Lanzhou Institute of Glaciology and Cryopedology, Chinese Academy of Sciences, No.4: 88-96.

Chen, Q. 1985. *Formation and characteristics of the debris flow at the Jiangjia Gully of Dongchuan in Yunnan* (in Chinese). Memoirs of Lanzhou Institute of Glaciology and Cryopedology, No.4: 70-79.

Deng, Y. 1985. *A preliminary approach to the geologic and geomorphologic effect of debris flow* (in Chinese). Memoirs of Lanzhou Institute of Glaciology and Cryopedology, No. 4: 241-250.

Du, R. & S. Zhang 1985. *A large-scale debris flow in the Guxiang Gully, Tibet in 1953* (in Chinese). Memoirs of Lanzhou Institute of Glaciology and Cryopedology, No. 4.

Du, R., Z. Kang & S. Zhang 1986. On the classification of debris flow in China. *Proc. of the Third Intern. Symp. on River Sedimentation*, pp. 1286-1292.

Janda, R.J., K.M. Scott, K.M. Nolan & H.A. Martinson 1986. *Lahar movement, effects and deposits.* US Geological Survey.

Freshman, C.M. 1978. *Debris flow.* Hydrology and Meteorology Press (in Russian).

Kang, Z. 1985a. *A velocity analysis of viscous debris flow at Jiangjia Gully of Dongchuan in Yunnan* (in Chinese). Memoirs of Lanzhou Ins. of Gla. and Cryo., No. 4.

Kang, Z. 1985b. *Characteristics of the flow patterns of debris flow at Jiangjia Gully in Yunnan* (in Chinese). Memoirs of Lanzhou Institute of Glaciology and Cryopedology, No. 4: 97-100.

Li, H. & Y. Deng 1985. *Distribution and deposit features of debris flow in China* (in Chinese). Memoirs of Lanzhou Institute of Glaciology and Cryopedology, No. 4:251-255.

Li, J. & D. Luo 1985. *Characteristics of debris flow in China and control engineering* (in Chinese). Bulletin of Soil and Water Conservation, No. 2: 15-18.

Pierson, T.C. 1986. Flow behavior of channelized debris flow, Mount St. Helens, Washington. In Abrahams, Boston, Allen & Unwin (eds), *Hillslope Processes*, pp. 269-296.

Qian, N. (Ning Chien) 1980. A comparison of the bed load formulas (in Chinese). *J. of Hyd. Engin.*, No. 4.

Qian, N. (N.Chien) & Z. Wang 1984. A preliminary study on the mechanism of debris flows (in Chinese). *Acta Geographica Sinica*, Vol. 39(1).

Sieyama, T & S. Woemoto 1981. Characteristics of debris flow at bends. *J. of Civil Eng.*, No. 5.

Takahashi, T. 1978. Mechanical characteristics of debris flow. *J. of Hydraulic Division, ASCE*, HY8: 1153-1169.

Takahashi, T. 1980. Debris flow on prismatic open channel. *J. of Hydraulic Engineering, ASCE*, 106(HY3): 381-196.

Wang, Z. 1987. Buoyancy force in solid-liquid mixtures. *Proc. of 22nd Cong. of IAHR, Lausanne, Switzerland.*

Wang, Z. & N. Qian (N. Chien) 1985a. A preliminary investigation on the mechanism of hyperconcentrated flow. *Proc. of Intern. Workshop on Flow at Hyperconcentration of Sediment, IRTCES*, pp. II4-1-16.

Wang, Z. & N. Qian (N. Chien) 1985b. Experimental study of motion of laminated load. *Scientia, Sinica, Ser. A*, Vol. 28(1): 102-112.

Wang, Z. & N. Qian (N. Chien) 1989. Characteristics and mechanism of debris flow. *Proc. 4th Symp. on River Sedimentation, IRTCES*, 722-729.

Wang, Z. & X. Zhang 1989. Experimental study of Initiation and laws of motion of debris flow (in Chinese). *Acta Geographica Sinica*, Vol. 44(3): 291-301.

Wang, Z. & X. Zhang 1990. Initiation and laws of motion of debris flow, Hydraulics/ Hydrology of Arid land (ed. R.H. French). Published by the American Society of Civil Engineers, pp. 596-601.

Wang, Z., J. Huang & Q. Zeng 1989. Structure of clay and its effects on open channel flow (in Chinese). *J. of Hyd. Engin.*

Wang, Z., B. Lin & X. Zhang 1989. Instability of non-Newtonian flow (in Chinese). *Acta Mechanica Sinica*, pp.266-275.

Zhang, S. & J. Yuan 1985. *Impact force of debris flow and its detection* (in Chinese). Memoirs of Lanzhou Institute of Glaciology and Cryopedology, No. 4.

Zhang, S., W. Yang & X. Chen 1985. *Formation and Characteristics of the debris flow along Dabai River section of Xiaojiang River in Yunnan* (in Chinese). Memoirs of Lanzhou Inst. of Glac. and Gryo., No. 4.

The utilization of hyperconcentrated flow

Hyperconcentrated flow is characterized by its large sediment-carrying capacity, as mentioned in Chapter 8. Such large sediment-carrying capacity has been utilized for different purposes in various fields. The utilization of hyperconcentrated flow will be illustrated in this chapter.

12.1 HYPERCONCENTRATED IRRIGATION/WARPING

12.1.1 *Development of hyperconcentrated irrigation/warping*

Hyperconcentrated warping is popular in North China, such as Hebei, Shanxi, Shannxi and other provinces. In Zhaolaoyu Irrigation District in Shannxi province utilizing hyperconcentrated flood for agricultural purpose started early in Qing Dynasty, that is, 2300 years ago. But in the past hyperconcentrated warping/irrigation developed only in small irrigation districts, where muddy water was directly diverted into farm land through canals of short distance.

In seventies of this century hyperconcentrated irrigation/warping rapidly developed in large irrigation districts such as Luohui, Jinghui and Baojixia Irrigation districts in Shannxi Province (Shannxi Research Group of Hyperconcentrated Irrigation/Warping, 1976). In the past in these districts flow at concentration higher than 167 kg/m³ was not allowed to be diverted under the consideration of avoiding serious siltation in canals.

The situation in these irrigation districts is similar in some respects. In all these irrigation districts water is taken from heavily sediment-laden rivers, that is, North Luohe River, Jinghe River and Weihe River. On these rivers the average concentration in flood season is from 200 kg/m³ to more than 300 kg/m³, the maximum concentration can be higher than 1000 kg/m³. The main parameters of incoming sediment of these river are illustrated in Table 12.1.

In history water diversion had to be interrupted during hyperconcentrated floods. For instance, according to the statistical data in the period of 1952-1974, in

Table 12.1. The situation of incoming sediment of North Luohe, Jinghe and Weihe River.

River	Gauging Station	Sediment		Maximum concentration (kg/m³)	Average concentration in flood season (kg/m³)	Suspended load	
		Annual load (10⁹t)	Percentage in flood season*			D_{50} (mm)	D_{cp} (mm)
Jinghe	Zhangjiashan	0.31	92.8%	1012	312	0.038	
North Luohe	Zhuangtou	0.095	90.0%	1093	345	0.0353-0.042	0.0399-0.0431
Weihe	Linjiachun	0.207	76.3%	840	189	0.02-0.026	0.025-0.0545
Weihe	Weijiapu	0.216		732		0.0189-0.025	0.0289-0.0483

*Ratio of incoming sediment in flood season to the total incoming sediment.

Table 12.2. Basic parameters of several main canals in Luohui irrigation districts.

Canal	Designed discharge (m³/s)	Slope (10⁻⁴)	Cross section		Boundary condition
			Depth (m)	Width of bottom (m)	
Great Main Canal	18.5	4	2	5	Stone lining and concrete lining alternatively
East Main Canal	5	3.3	1.4-4.8	2.2	Concrete lining
Middle Main Canal	3	4-5	0.86-1.22	2.5-5.0	Unlining
West Main Canal before reconstruction	3	4-6.67	1.64	1.5-2.0	Unlining
West Main Canal after reconstruction	5	4	1.58	1.8	Concrete lining
Luo West Main Canal	5.5	3.64	1.85	1.6	Concrete lining
Luo West Main Canal No. 1	5	3.64	1.85	1.6	Concrete lining with frequent siltation
Middle 2nd lateral canal	0.5	6.67-10	0.65	0.8	Unlining
Middle 2nd lateral by-canal	0.5	10	0.65	0.8	Concrete lining canal with disk shape

North Luohe River more than 350 hours annually the concentration of incoming flow is higher than 167 kg/m³. In these irrigation districts water supply is not sufficient, particlarly in Summer time, when the temperature is high and vaporization is strong. The interruption of water diversion brought agriculture heavy loss. Besides, saline-alkaline land and depression, which need to be improved, widely spread in these irrigation districts. On the other hand, compared with that in the lower reach of the Yellow River, the land surface in the middle reach of the Yellow River has a steeper slope. Correspondingly, canals in the middle reach can be arranged with a steeper slope, which favors the transportation of hyperconcentrated flow.

Basic parameters of several main canals in Luohui Irrigation District are listed in Table 12.2 (Xu & Wan, 1985).

Through practise and systematic research and field observation people accumulated experience and mastered mechanism of hyperconcentrated flow, and

Table 12.3. The situation of hyperconcentrated irrigation/warping in the period of 1974-1976.

Irrigation district	Jinghui	Luohui	Baojixia	
			Upper part	Lower part
Maximum concentration diverted (kg/m³)	450	949	448	560
Annual sediment amount diverted in flood season (10³t)	8160	7690	7000	3270

hyperconcentrated irrigation/warping developed rapidly in seventies. More water and sediment was diverted and utilized for agriculture purpose. It can be reflected in Table 12.3.

12.2.2 *Criteria for siltation judgement*

During the development of hyperconcentrated irrigation/warping some criteria for siltation judgement have been developed in different irrigation districts based on local conditions.

It can be seen from Figure 8.8 in Chapter 8, data from irrigation districts also follow the hook-like sediment-carrying capacity correlationship. By further analysing factors contained in the correlationship, it was found that under the condition of canals, water depth H varied in a narrow range and there was a rather constant correlationship between concentration and settling velocity W_0 (or corresponding D_{50}) of incoming sediment. Therefore among these factors flow velocity V becomes the most important one, playing a predominant role. Based on such analysis, a criterion for siltation judgement was established in Baojixia Irrigation District, see Figure 12.1 (Shannxi Research Group, 1976). Here V is the average flow velocity in canal. S_w is the concentration in weight percentage, that is, the ratio between the sediment weight and the weight of muddy water in unit volume. It was found that data from Luohui Irrigation District also obey the criterion.

Based on field data in Luohui Irrigation District the critical velocity for siltation under different conditions was ascertained as follows:
Under the condition of $S_w = 20$-30%

$$V_{cr} = 0.7\text{-}0.8 \text{ m/s for main canals } (H = 1 \text{ m})$$

$$V_{cr} = 0.6\text{-}0.7 \text{ m/s for lateral canals } (H = 0.3\text{-}0.4 \text{ m})$$

or

$$\left(\frac{V^3}{H}\right)_{cr} = 0.4\text{-}0.5$$

Under the condition of $S_w > 50\%$

$$V_{cr} = 0.87\text{-}0.9 \text{ m/s}$$

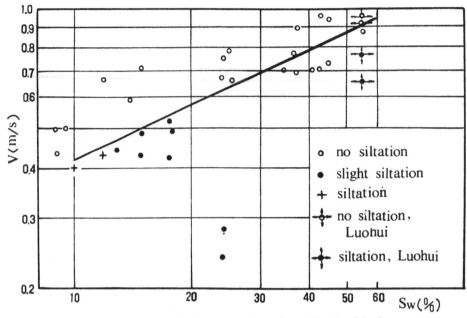

Figure 12.1. A criterion for siltation judgement in Baojixia Irrigation District.

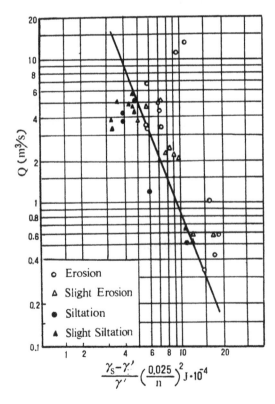

Figure 12.2. A criterion for siltation judgement in Luohui Irrigation District.

or

$$\left(\frac{V^3}{H}\right)_{cr} = 0.6$$

For planning and design work velocity is not a proper parameter, and discharge Q and slope J are more convenient to be used. A criterion containing Q and J has been established as shown in Figure 12.2. The criterion can be written as: (Wan & Xu, 1984)

$$Q = 1.86\times10^{-8}\left[\frac{\gamma_s - \gamma_m}{\gamma_m}\left(\frac{0.025}{n}\right)^2 J\right]^{-2.56} \tag{12.1}$$

in which Q is in m^3/s, γ_s and γ_m, specific weight of sediment and that of muddy water, respectively, n Manning coefficient of the canal.

12.1.3 *Experience in design and management of hyperconcentrated irrigation/warping*

Through practice and research work some experience and knowledge on hyperconcentrated irrigation/warping have been accumulated.

In design of hyperconcentrated irrigation/warping canal system, the following points should be noticed: (Luohui Research Group of hyperconcentrated irrigation/warping, 1978)

1. It is not necessary to absolutely avoid siltation in canals during hyperconcentrated irrigation/warping. What one should do is to keep essentially equilibrium in an irrigation season or in a year. It is very difficult to keep canals completely 'clean' during hyperconcentrated irrigation/warping. If one asked for it, the area of hyperconcentrated irrigation/warping would have been limited or greatly reduced. In most cases siltation in canals is inevitable. But if the siltation is not serious and a certain discharge can still be conveyed through canals, the normal irrigation will not be interfered too much and sediment deposited during hyperconcentrated irrigation/warping will be eroded gradually when the concentration of incoming flow drops and normal irrigation continues.

This situation should be taken into account in design. That is, an enough freeboard of canals should be designed in advance.

2. Canal with narrow and deep cross section favors the transport of hyperconcentrated flow. The Middle Main Canal of Louhui Irrigation District is a unlining canal with weed growing along its banks. During hyperconcentrated irrigation/warping larger than usual discharge passes through the canal and weed is submerged in muddy water. Flow near banks slows down, sediment deposits and buries weed on banks and two nearly vertical smooth walls are formed. But flow in central part of the canal keeps its high velocity and large amount of sediment can still be transported. The self-adjustment of canals during hyperconcentrated irrigation/warping reveals us with some idea. In Luohui Irrigation District during

the reconstruction period (1974-1975) all the rebuilt canals were designed as narrow and deep ones. Combined with other measures,rebuilt canals work well.

3. Canals should be lined if it is possible.

A lining canal has smaller Manning Roughness. Besides, during the erosion period after hyperconcentrated irrigation/warping in a lining canal previous deposits slides along the concrete surface and the canal remains regular in form. But in a unlining canal irregular erosion makes the canal a jagged, interlocking pattern.

4. Canals should have steep slope if local topographic condition allows. In Luohui Irrigation District during the reconstruction period some waterfalls were cancealed and some bridges and culverts causing backwater effects were rebuilt. By this way the effective slope of canals was raised.

5. Sluiceways should be installed at proper places.

By openning sluicegate local slope of water surface rises. It favors the transport of hyperconcentrated flow.

In irrigation management the following experience could be referred.

1. Hyperconcentrated irrigation/warping should be carried out with large discharge.

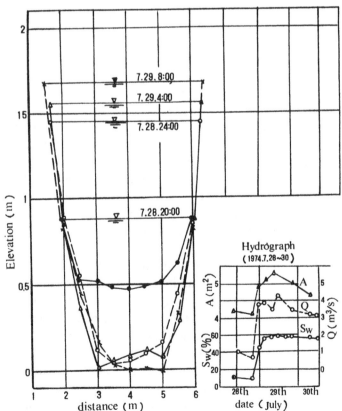

Figure 12.3. Variation of cross section, Middle Main Channel of Luohui Irrigation District.

In Luohui Irrigation District usually the incoming discharge is distributed into three main canals, that is, East Main Canal, Middle Main Canal and West Main Canal. During hyperconcentrated irrigation/warping the entire discharge is concentrated into one of these three canals and other two canals are closed. And these three canals are used in turn in a flood season. In Figure 12.3 is the variation of a cross section of the Middle Main Canal. Before hyperconcentrated warping the discharge was less than 1 m³/s. The discharge was raised up to more than 4 m³/s when hyperconcentrated warping was started. Although the concentration increased to S_w = 40-50%, the canal was eroded, instead of silted.

In Luohui Irrigation District for each individual main canal hyperconcentrated irrigation/warping is carried out with large discharge, and siltation can be effectively avoided or greatly reduced. Normal irrigation is carried out with small discharge, and erosion under the condition of low concentration can be avoided, too.

2. Canals should be well maintained. Particularly, weed growing along banks of unlining canals should be cleaned regularly. Under the condition of good maintenance the Manning roughness of canal can be kept in a low value, and it favors the transport of hyperconcentrated flow.

3. In case canals are silted during hyperconcentrated irrigation/warping, stirring deposit by machine or man power is an effective way of keeping canals clean. Since hyperconcentrated slurry behaves as time-dependent fluid (thixotropic fluid), strong stirring fluidizes it and original stagnation layer starts flowing again. The stirred slurry can be carried away by the current. It is a common situation that in Luohui Irrigation District during hyperconcentrated irrigation/warping local farmers together with their cattle and tractors work hard along canals to stir the stagnation layer of slurry and fluidize it again.

12.1.4 *The effect of hyperconcentrated irrigation/warping*

Hyperconcentrated irrigation/warping is popular in Shannxi Province. Advantage of hyperconcentrated irrigation/warping can be summarized as follows.

1. Water suppply in these districts is increased by carring out hyperconcentrated irrigation (Xu et al., 1989).

Water supply is insufficient in these districts and the situation is particularly severe in July and August, when most hyperconcentrated floods happen. Hyperconcentrated irrigation eases the tension to a certain degree and great advantage of raising agricultural production is obtained. According to statistics, in the period of 1969-1985, 2.06×10^8 m³ of turbid water, which corresponds to one fifth of the total diverted water volume in summer, was diverted. Correspondingly, 1.06×10^8 tons of sediment, which amounts to 9.8% of the sediment load of the North Luohui River, was diverted and utilized. The diverted water volume and other indexes involving hyperconcentrated irrigation/warping in Luohui Irrigation District are listed in Table 12.4.

Table 12.4. The main situation of hyperconcentrated warping and irrigation in Luohui Irrigation District (1969-1981).

Year	Diverted water volume* ($10^6 m^3$)	Diverted sediment ($10^6 t$)	Maximum concentration (kg/m^3)	Area of hyperconcentrated irrigation (hectares)	Warping area (hectares)	Improved saline land** (hectares)
1969	4.07	2.40	818.2	1195		
1970	16.8	8.46	793.8	9449	439	
1971	12.9	8.00	940.0	5022	567	
1972	12.5	8.94	913.6	4816	435	
1973	20.1	14.20	884.4	11291	701	
1974	9.71	7.69	952.9	2086	352	
1975	1.12	0.53	482.0			
1976	6.21	3.96	629.8	2058	105	1333***
1977	24.9	15.1	946.3	7789	1170	545
1978	19.2	9.73	916.1	5918	843	393
1979	20.8	11.5	954.0	7038	1008	657
1980	12.3	9.28	916.1	4990	634	445
1981	8.87	6.60	949.2	2962	481	350
	169.5	106.4	964.3	64616	6735	3723

*Only water with concentration higher than $167 \, kg/m^3$ is counted.
**Improved saline land is included in warping area.
***Accumulated area of 1969-1976.

Table 12.5. The increase of agricultural production due to hyperconcentrated irrigation.

	Length of ear (cm)	Percentage of empty ears (%)	Weight of an ear (g)	Weight of one thousand grains	Output per hectare (kg)
Without hyper. irrig.	19.61	11.1	360	204.5	3000
With hyper. irrig.	19.85	7.9	400	221.8	3675

An example showing the increase of agricultural production due to hyperconcentrated irrigation is listed in Table 12.5, in which the situation of corn production in two neighbouring plots, one with hyperconcentrated irrigation and one without it, is compared.

It shows that the output was raised by means of hyperconcentrated irrigation.

In some cases when the drought is severe plant dies and no harvest can be obtained provided hyperconcentrated irrigation could not be carried out.

2. Soil constituent can be improved by hyperconcentrated warping.

There are some heavy clay land and some sand land in these irrigation districts. The top soil layer of such land is lack of granular aggregates and little water can be held in it. Such land is not suitable for cultivation. Sediment carried by hyperconcentrated flow is mostly constituted of silt and fine sand with a small amount of clay. The deposition forms good soil suitable for cultivation. Some examples of soil improvement are shown in Table 12.6.

3. Salt in saline land can be washed away by hyperconcentrated warping.

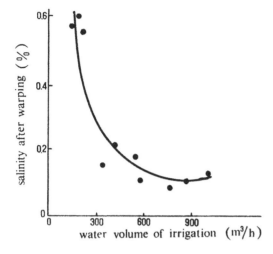

Figure 12.4. The correlationship between the water volume used for warping and the salinity in top soil after warping.

Table 12.6. The variation of the gradation composition of top layers after hyperconcentrated warping.

Place	Depth from the ground (cm)	Median diameter (mm)		Percentage of particles finer than 0.01 mm (%)			Remark
		Before warping	After warping	Before warping	After warping	Increment	
Cheng Zhuang	0-10	0.034	0.043	20.3	16.6		1978
	10-20	0.032	0.039	23.9	21.5		
	20-30	0.017	0.035	36.2	22.8		
	Average	0.028	0.039	26.8	20.8	−24.3	
Luihui Farmland	0-10	0.007	0.046	72.4	14.5		1979
	10-20	0.008	0.044	55.0	16.2		
	20-30	0.014	0.048	42.2	14.3		
	Average	0.010	0.046	56.5	15.0	−73.5	

During hyperconcentrated flood plenty of water can be used for warping. The more the water used, the better the desalinization. Based on field data in Luohui Irrigation District, a correlationship between the water volume used for warping and the salinity in top soil after warping is shown in Figure 12.4.

Besides, hyperconcentrated warping is always carried out in summer when the temperature is high. And desalinization can get better effect in summer when the temperature is high, particularly for sulphate. The solubility of sulphate of sodium is 48.5 g/l at 0°C and 410 g/l at 40°C. As a result, the effect of desalinization by hyperconcentrated warping is extremely good.

Before hyperconcentrated warping more salt is concentrated in the top soil layer due to intensive vaporation (upper plot of Figure 12.5). After warping quite a lot salt has been washed away and less salinity exists in the top layer (lower plot of Figure 12.5). Such change obviously favors the growth of plant.

It should be emphasized that the effect of hyperconcentrated warping can be

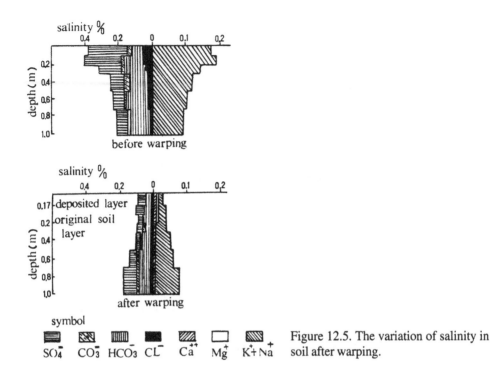

Figure 12.5. The variation of salinity in soil after warping.

maintained only if the drainage system there is profound.

4. The fertility of soil can be raised by hyperconcentrated warping.

Hyperconcentrated floods always carry large amount of decomposed leaves and branches, excrement and urine of cattle, etc. Some organic fertilizer is absorbed on the surface of sediment grains, particularly, on the surface of fine grains. Based on field measurement in Luohui Irrigation District the content of full nitrogen, full phosphorus, quick-acting nitrogen, quick-acting phosphorus and organic matter are 0.058%, 0.052%, 75 ppm, 7.5 ppm and 0.929%, respectively. Based on field data collected in Yangjiazhuang Village, Luohui Irrigation District in 1977, after hyperconcentrated warping the increase of fertility is obvious as shown in Table 12.7.

Due to the combined effect of all above mentioned factors, the agricultural production can be raised significantly by hyperconcentrated warping. Some typical examples in Luohui Irrigation District are shown in Table 12.8.

5. Sediment entering the lower Yellow River can be reduced by hyperconcentrated irrigation/warping.

Sediment causing the aggradation of the Lower Yellow River originates from the loess plateau in its middle reaches. Hyperconcentrated irrigation/warping carried out in the area of its middle reaches reduces the sediment amount entering the Lower Yellow River. Sediment which is a heavy load to the Lower Yellow River turns to be resources utilized in its middle reach area. Based on statistics in

Table 12.7. The variation of fertilizer content.

Place	Depth (cm)	Organic matter (%)			Full nitrogen (%)			Full phosphorus		
		Before warping	After warping	% in-crease	Before warping	After warping	% in-crease	Before warping	After warping	% in-crease
Near the inlet	0-10	0.4952	0.5995		0.0229	0.0442		0.0540	0.0620	
	10-20	0.3827	0.5995		0.0179	0.0351		0.0515	0.0670	
	Average	0.4390	0.5995	36.6	0.0204	0.0391	94.6	0.0525	0.0645	22.9
Near the end	0-10	0.4727	0.7194		0.0286	0.0457		0.0510	0.0600	
	10-20	0.4952	0.6714		0.0274	0.0373		0.0520	0.0600	
	Average	0.4840	0.6954	43.7	0.0280	0.0415	48.2	0.0515	0.0600	16.9

Table 12.8. Raising grain production by hyperconcentrated warping.

Place	Time	Depth of deposition (cm)	Plant	Area (hectare)	Grain output (kg/hec.)	
					Before warping	After warping
Shonglu, Dali	1971.7	35	Wheat	2.67	0	1875
Yangjiazhong, Dali	1977.7	25	Wheat	4.0	0	2758
Fuxin, Pucheng	1977.7	20	Barley	2.67	0	1800
Fuxin, Pucheng	1978.7	30	Wheat	2.67	0	1481
Jianji, Pucheng	1979.7	69	Wheat	8.13	0	2419
Mingdi, Pucheng	1979.7	26	Wheat	1.73	1950	3900
Mingdi, Pucheng	1979.7	26	Rape	1.52	1950 (wheat)	1875
Liohui, Farm	1979.7	32	Wheat	2.95	0	2782
Mingdi, Pucheng	1979.7	30	Wheat	3.73	0	2782
712 farm	1979.7	30	Wheat	3.40	0	3608
Mingdi, Pucheng	1979.7	45	Wheat	1.0	0	2168
Tongyi, Pucheng	1979.7	30	Wheat	1.33	0	2452

1974-1976, the amount of sediment utilized in these districts is still limited. But if the measure of hyperconcentrated irrigation/warping is widely popularized in the middle reaches, the reduction of sediment may reach a certain percentage.

12.2 RELEASING SEDIMENT FROM RESERVOIRS IN THE FORM OF HYPERCONCENTRATED FLOW

The powerful sediment-carrying capacity, which is one of the distinct character-istics of hyperconcentrated flow, can be utilized for releasing sediment from reservoir.

As described in Chapter 10, entering a reserevoir, a hyperconcentrated flow can easily submerge underneath the clear water and form a density current due to its large difference in density. Hyperconcentrated density current moves at higher velocity than a common density current and reaches the dam in a shorter time. Due to the high viscosity of the hyperconcentrated flow very few or even no particles settle during the motion of hyperconcentrated density current, and the mixing at the interface between the clear water and the turbid water is weak too. Reaching the dam, the hyperconcentrated fluid is partly stored there and a subreservoir of hyperconcentrated fluid will be formed underneath the clear water provided the releasing discharge is less than the incoming one. Sediment in the subreservoir settles very slowly and the turbid water keeps its fluidity for rather long time. Provided the bottom sluice is opened, the releasing of turbid water may last for a couple of days. As the result of all these afore-mentioned favourable factors, the releasing rate, that is the ratio of released sediment to the incoming sediment, can be rather high.

Take Heisonglin Reservoir in Shannxi Province as an example (Ren & Zhao, 1973). There systematic field measurement has been carried out, and main parameters of seven recorded events of hyperconcentrated density currents are listed in Table 12.9. It can be seen from the table that the releasing rate for a hyperconcentrated density current reaches 50% or even more. As a summary, the total incoming sediment carried by those density current amounts to 0.95 million tons, of which 0.58 million tons is released. The average releasing ratio reaches 61.2%, with a maximum releasing ratio of 91.4%.

Liujiaxia is a large reservoir located on the stem of the Upper Yellow River. A concrete dam of more than 140 m height is located in a narrow gorge and a reservoir with a storage capacity of 5.7 billion cubic meter is formed. One of its

Table 12.9. Hyperconcentrated density currents in Heisonglin Reservoir.

Time (y.m.d.)	Incoming flow				Outcoming flow				Highest stage in front of the dam (m)	Releasing ratio (%)
	Peak discharge (cu.m/s)	Maxim. concentration (kg/cu.m)	Water volume (10^4 cu.m)	Amount of sediment (10^4t)	Maxim. discharge (cu.m/s)	Maxim. concentration (kg/cu.m)	Water volume (10^4 cu.m)	Amount of sediment (10^4t)		
64.7.11	132.0	534	88	23.79	4.8	582	77.9	11.69	753.8	37.9
64.7.16	35.0	224	57.4	7.88	7.3	338	62.9	6.72	751.2	66.0
64.8.1	135.0	731	72.1	26.89	7.0	749	83.5	30.88	751.4	91.4
65.7.19	23.6	389	50.0	6.135	6.5	573	37.15	2.83	753.36	35.7
66.8.9	23.2	472	15.2	5.13	4.6	340	18.9	2.56	752.166	50.0
71.7.21	18.1	456	13.4	3.09	3.96	227.2	23.65	1.825	754.948	59.0
72.8.1	30.9	430	10.67	3.46	5.10	273.0	16.08	1.867	753.82	54.0

Table 12.10. Recorded density currents in Liujiaxia Reservoir in 1974.

Date	Incoming sediment (10^4t)	Released sediment (10^4t)	Releasing ratio (%)
July 4-5	12.0	3.98	33.1
July 22-23	55.2	21.0	38.1
July 25-28	176.0	119.7	68.0
July 31	107.0	26.5	24.8
Aug. 14	45.1	2.43	5.4

tributaries, Taohe River, confluences at a point 2 km upstream from the dam site. Upstream from confluence, Taohe River runs in a narrow gorge with a steep slope I (0.0025-0.010). Hyperconcentrated floods from the Taohe River easily form density currents and reach the dam site. Some of the recorded density currents are listed in Table 12.10.

Although the concentration of Taohe River is not very high, the releasing ratio is still rather high. Besides, as an overall tendency, the larger the amount of incoming sediment, the larger the releasing ratio. The reason for larger releasing ratio is that higher concentration associated with larger amount of incoming sediment makes density current flow more fast, deposit less coarse particles. Consequently, more sediment can be sluiced.

In order to release sediment from the reservoir with the minimum amount of water, properly operating sluice gates is important. Sluice gates should be just opened as soon as density current arrives at the damsite. For this purpose an empirical relationship shown in Figure 12.6 is established based on field data. In

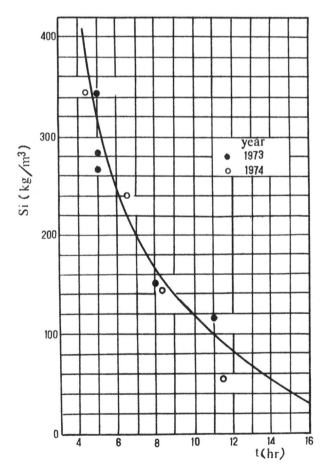

Figure 12.6. The correlationship between travel time t and concentration S_i in Liujiaxia Reservoir.

the figure S_i is the concentration of incoming flow, t is the travel time of the density current, in hour.

It is clear that the higher the concentration of incoming flow, the shorter the travel time. Under the concrete conditions of Liujiaxia resevoir Taohe River runs through a narrow gorge with steep slope, the variation of the water level in front of the dam does not make the cross section area and the length of reservoir change too much. Therefore, the water level in front of the dam does not have obvious influence on the travel time of density current.

Density current is not the only way of releasing sediment from reservoirs. Many reservoirs are operated according to the principle of 'storing clear water in non-flood season and discharging turbid water in flood season'. In flood season water level in such reservoirs is low and they actually behave as alluvial rivers. Hyperconcentrated floods just pass through reservoirs in the form of open channel flow. Little or no deposition is associated with hyperconcentrated floods. In some cases erosion happens and the releasing ratio is even larger than 100%. Hyperconcentrated floods passing through Sanmenxia Reservoir in 1977 are such examples. Main parameters of these floods are listed in Table 1.3. Other examples are recorded hyperconcentrated floods passing through Heisonglin Reservoir, shown in Table 12.11. All the examples show the high releasing ratio of the hyperconcentrated flow. One important point in the operation of reservoir is to maintain a low water level for keeping the flood in the main channel. Otherwise once hyperconcentrated flood overflows on flood plain, serious siltation on flood plain will be inevitable and the storage capacity of the reservoir drops rapidly.

In some arid area where water is extremely cherished, even turbid water coming in flood season should not be released without utilization. In the meantime, the storage capacity of the reservoir should be kept too. Under such condition the operation scheme of 'discharging turbid water in flood season and storing clear water in non-flood season' should be modified in order to meet the

Table 12.11. Hyperconcentrated open channel flow in Heisonglin Reservoir.

Time (y.m.d.)	Incoming flow				Outcoming flow				Highest stage in front of the dam (m)	Releasing ratio (%)
	Peak discharge (cu.m/s)	Maxim. concentration (kg/cu.m)	Water volume (10^4 cu.m)	Amount of sediment (10^4t)	Maxim. discharge (cu.m/s)	Maxim. concentration (kg/cu.m)	Water volume (10^4 cu.m)	Amount of sediment (10^4t)		
63.8.8	44.2	364	23.53	4.412	6.39	598	32.34	7.60	747.0	133.2
68.8.11	89.0	564	103.36	35.865	10.72	508	122.8	47.425	748.1	102.2
64.8.10	92.2	427	83.05	24.34	7.6	540	93.0	20.72	748.05	82.2
69.8.9	215.0	585	321.00	100.0	6.54	418	247.9	38.95	756.646	39.0
70.7.24	63.4	415	41.4	8.35	5.8	514	38.9	7.47	749.1	90.0
70.8.4	370.0	438	326.0	118.0	8.79	606	255.00	75.76	757.226	64.2
71.8.19	19.3	369	14.5	2.92	4.54	403	13.05	3.13	748.15	108.0
71.8.20	465.0	314	592.7	148.4	10.75	621	488.00	86.20	761.982	58.0
72.8.16	21.3	774.4	12.01	4.9	3.22	630	15.38	4.09	746.4	84.0

harsh requirement. Experience and operation scheme developed in Hengshan Reservoir might be a good example of dealing with the contradiction between the water supply and the preservation of storage capacity (Zhou et al., 1989).

Hengshan Reservoir is a reservoir of median scale with a storage capacity of 13.1×10^6 m^3. The height of its dam is 69 m. A flood-discharge tunnel with a discharge capacity of 1260 m^3/s and a bottom sluice with a discharge capacity of 17 m^3/s are located 14.5 m and 2.6 m above the bed, respectively. The original river bed with a slope of 0.029 consisted of gravel and pebbles. The annual runoff of 1.33×10^7 m^3 and the annual sediment load of 753×10^3 t are concentrated in the flood season. Median diameter of incoming sediment varies in the range of 0.011-0.058 mm. The average concentration of floods is 385 kg/m^3 with a peak value of 836 kg/m^3. Of the total storage capacity, 25.3% was lost during the impounding stage in 1966-1973, when the thickness of deposition in front of the dam was 27 m. In July and August of 1974, the reservoir was emptied for 53 days. A total of 1.02×10^6 m^3 of sediment, including the original deposit of 7.16×10^5 m^3 and the incoming sediment for that year was released, with 2.6 m^3 of water being consumed for sluicing 1 t of sediment. Based on the aforementioned practice, the principle of the new operation scheme was established as follows: 'Storing water and sluicing sediment in normal years, emptying the reservoir while releasing sediment for downstream irrigation and warping', that is, impounding water for 3-5 years and then emptying the reservoir, waiting for a flood passing through it and eroding the deposit.

Table 12.12 shows the history of the operation of Hengshan Reservoir. Before

Table 12.12. The operation of Hengshan Reservoir.

Year	Opera-tion	Status of S_i	S_i (10^{3}m^3)	S_o (10^{3}m^3)	S_o/S_i (%)	W_o/S_o (m^3/t)	Empty days	S_i-S_o (10^{3}m^3)	S (10^{3}m^3)
66-67	A	2 abun	2433	1540	63.3		54	+ 893	893
68-73		1 abun, 2 mean, 3 dry	3629	1330	36.3			+2299	3192
74	E	mean	304	1020	335.5	2.63	53	− 716	2576
75-78	I	2 abun, 1 mean, 1 dry	1921	960	50.5			+ 961	3437
79	E	mean	432	1250	289.4	9.3	52	− 818	2620
80-81	I	2 mean	733	462	63.0			+ 271	2891
82	E	abun	629	1426	226.7	5.65	56	− 797	2094
83-84	I	2 dry	538	185	34.4			+ 353	2447

Note: St. = Stage; A = Reservoir emptying before flood season and impounding during flood period; I = impounding during floods and sluicing sediment; E = Emptying reservoir and sluicing sediment; 2 abun, 1 mean, 1 dry = There were two abundent sediment years, one mean year, and one dry year in this stage; S_i, S_o = Incoming and sluiced off sediment load, respectively; S_i-S_o = Amount of sediment deposited in reservoir; S_o/S_i = Ratio of sluiced off sediment load to incoming sediment load; W_o/S_o = Water used for slucing one ton of sediment; S = Accumulated mass amount of sedimentation.

1973 (Stage I) the reservoir was impounded in flood season and it was silted continuously. Since 1974 (Stage II) the operation scheme has been changed from impoundment into perennial regulation combined with emptying reservoir occasionally. And it was alternately silted or eroded in stage II with a total sedimentation ranging from $2.09-3.02\times10^6$m^3. From a long-term view the reservoir is nearly in equilibrium.

The operation scheme of regulating water and sediment in stage II possesses following characteristics:

1. Regulating sediment over years by using the storage capacity of the main channel and maintaining the long-term storage capacity in a time. 5.73×10^6m^3 of sediment, which was more than the incoming sediment, was sluiced in 1974-1986. But the overall releasing ratio in stage I before 1973 was less than 50%.

2. Sluicing sediment in the form of hyperconcentrated flow and with small amounts of water. For sluicing 1 t sediment 3-10 m^3 of water was consumed in the impounding stage, but only 2-5 m^3 of water was consumed in stage II. The incoming sediment of 3-5 years was sluiced in a flood season by 7% of the total incoming water. Compared with the situation before 1973, a large amount of water was saved.

3. Making full use of water and sediment resources.

During the impounding period stage II, entering the reservoir, hyperconcentrated floods formed density current and then a subreservoir in front of the dam. Releasing sediment lasted for rather long time and more than 90% of the released sediment, which contained a certain amount of organic fertilizer, was diverted into the irrigation district and utilized. Except that part deposited on the flood plain, 60% of the incoming sediment was diverted into the irrigation district and utilized.

4. Developing fishery in the 3-5 year impounding period.

12.3 THE IDEA OF TRANSPORTING SEDIMENT OF THE YELLOW RIVER INTO SEA BY HYPERCONCENTRATED FLOW

The Yellow River is famous for its tremendous amount of sediment. According to historical statistics, the annual sediment load reaches 1.6 billion tons and about one fourth is deposited along its Lower Reaches. Big problem is caused by the deposition of sediment and the aggradation of the Lower Yellow River. The aggradation of Lower Yellow River is caused essentially by particles coarser than 0.05 mm, which is mainly carried by hyperconcentrated floods, originating from its Middle Reach area. According to the analysis of 103 floods from the years 1952-1960 before the construction of the Sanmenxia Reservoir and the years 1969-1978 after the completion of the dam, thirteen hyperconcentrated floods contribute to 60% of the total amount of deposition caused by the 103 floods (Qian et al., 1980). Although water-soil conservation work has been widely carried out in its Middle Reach area, the reduction of oncoming sediment is limited. On the other hand, with the development of agriculture and industry more and more clear water is diverted and utilized. The Yellow River faces the problem of water shortage both for economical development and for transporting sediment into sea. And the shortage is getting more and more serious. To solve this problem somebody suggested that the great sediment-carrying capacity of hyperconcentrated flow be utilized, that is, sediment be transported into sea by natural or artificially produced hyperconcentrated flow and saved clear water be utilized for agricultural or industrial purpose (Qi, 1989). As mentioned in Chapter 9, flow at concentration of $800 \, kg/m^3$ can be conveyed without deposition along rivers with narrow and deep channels under the condition of unit discharge larger than 5 m^3/s-m and longitudinal slope less than 10^{-4}. The longitudinal slope of the Lower Yellow River is about 2×10^{-4} in Henan Province and 10^{-4} in Shandong Province. It means the Lower Yellow River can meet the basic requirement of conveying the hyperconcentrated flow. But to realize the suggestion a series of difficulty are to be overcomed and lots of research work are needed to be done. First of all, incoming runoff and sediment load should be regulated in a set of reservoirs, such as Sanmenxia Reservoir and Xiaolangdi Reservoir, to produce hyperconcentrated flow as one expects. Secondly, the main channel of the Lower Yellow River should be transformed into a narrow and deep one and then it should be well kept. It is not easy to manage this. Now special study on these problems is being conducted.

12.4 HYDROTRANSPORT IN THE FORM OF HYPERCONCENTRATED FLOW

Hydrotransport is an important field in which the theory of hyperconcentrated flow can be used. Compared with hydrotransport at low concentration, hydrotransport at hyperconcentration has some advantages. Firstly, transporting solid granular material at high concentration can save water and energy in most cases. Secondly, these dense concentration mixtures settle without size segregation (i.e. they are 'stable'),and can be allowed to remain stationary in pipelines for long periods, and can easily be withdrawn after undisturbed storage in large vessels after several months (Elliott & Gliddon, 1970).

A proper design of the grain composition of transported granular material is important for energy consumption. Charles et al. (1971) found that the energy consumption could be saved by 8% provided the fine sand of 0.22 mm was transported by slurry consisting of 19% of clay and water, instead of pure water. But in Kenchington's (1978) experiments adding sand in clay slurry made pressure gradient reduce gradually. Generally speaking, granular material with wide range of size distribution is easier to be transported. In other words, less energy is required for transporting such material. Elliott also reported that pH value of a slurry has remarkable influence on its flow properties and consequent energy consumption in transport. All these phenomena can be illustrated by means of concept deduced in Chapter 6 and Chapter 8.

As mentioned in Chapter 6, the friction character of a pseudo-one-phase flow (or homogeneous flow) can be described by λ-Re_1 curve as shown in Figure 6.13 (Song & Wan, 1987). Here

$$\lambda = \frac{8gRJ_m}{U^2} \tag{12.2}$$

and

$$Re_1 = \frac{4\rho_m UR}{\eta\left(1 + \dfrac{2\tau_B R}{3\eta U}\right)} \tag{12.3}$$

in which ρ_m density of slurry, τ_B Bingham yield stress, η rigidity, U average velocity, R hydraulic radius and J_m energy gradient, in slurry column.

If a revised Reynolds number Re_1 consisting of both Bingham yield stress τ_B and rigidity η is used for pseudo-one-phase flow, the λ-Re_1 correlationship for a pseudo-one-phase flow has the same form as that for clear water flow. That is,

$$\lambda = \frac{64}{Re_1}, \qquad \text{for } Re_1 < 2300 \tag{12.4}$$

$$\lambda = \frac{0.316}{Re_1^{1/4}}, \quad \text{for } Re_1 > 2300 \tag{12.5}$$

And it tends to be horizontal line for $Re_1 > 5{\times}10^4$. Only in transitional region λ for pseudo-one-phase flow is a little lower than the corresponding λ for clear water flow.

The afore-mentioned formulas, including fluid properties ρ_m, τ_B, η, flow parameters U, J, as well as the pipe size R, have wide popularity.

In the engineering practise of pipeline design J-U curves are widely used. Actually, it is easy to deduce J_m-U curves from λ-Re_1 relationship. Here J_m is the energy slope in slurry column. In laminar flow region, the following equation can be deduced from Equations (12.2), (12.3) and (12.4),

$$J_m = \frac{2\eta}{\rho_m gR^2}\left(U + \frac{2\tau_B R}{3\eta}\right) \tag{12.6}$$

In hydraulically smooth flow (turbulent) region, based on Equations (12.2), (12.3) and (12.5), J_m can be expressed as:

$$J_m = \frac{0.04}{gR}\left(\frac{\eta}{4\rho_m R}\left(1 + \frac{2\tau_B R}{3\eta U}\right)\right)^{0.25} U^{1.75} \tag{12.7}$$

For a fully developed turbulent flow λ is a constant and the following equation can be written down:

$$J_m = \frac{k}{8gR} U^2 \tag{12.8}$$

in which k is a constant, depending on the relative roughness of the boundary.

Under the condition of given R, ρ_m, τ_B and η, all the three Equations (12.6), (12.7) and (12.8) reduce to J_m-U relationship.

Actually, under different condition two kinds of J_m-U relationships can be easily deduced from λ-Re_1 curves.

Under the condition of low velocity and rather high concentration of fine particles, the rigidity η and yield stress τ_B of transported slurry are quite large, the flow will be laminar and the J_m-U relationship can be described by Equation (12.6). In low velocity region, the first term in the brackets is much smaller than the second term, that is, $U \ll 2\,\tau_B R/3\eta$. Therefore, it can be neglected and we obtained:

$$J_m = \frac{4\tau_B}{3\rho_m gR} \tag{12.9}$$

It means that in low velocity region U has little influence on the pressure slope J_m, and the latter can be considered as a constant depending on properties of slurry (ρ_m, τ_B) and the size of pipe R. With U decreasing the J_m-U relationship tends to be a horizontal straight line as shown in Figure 12.7. In the figure a set of curves,

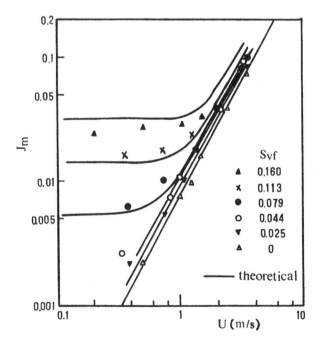

Figure 12.7. The relationship between J_m and U of suspension.

corresponding to several different slurries (or different τ_B, η), and some experiment dots are plotted. The curves deduced from λ-Re_1 relationship coincide with experiment dots quite well.

Under the condition of high velocity and low concentration sediment particles move as suspended load and the friction loss is the same as that of pure clear water, that is, the J_m-U curve of the sediment-laden flow is the same as that of clear water. With velocity decreasing bed load and consequent additional energy loss appear. The J_m-U relationship of a sediment-laden flow starts deviating from the J-U relationship of a clear water flow. With velocity further decreasing sediment particles deposit on bottom and bed configuration forms. Consequently, additional friction loss greatly increases and J_m-U relationship like that in Figure 12.8 appears.

The effect of pipe diameter on pressure gradient is closely concerned by design engineers. This problem is particularly important in transforming experiment results obtained in small pipes to prototype. There is no theoretical solution about this problem up to now. Fei has given some J-D relationships by comparing design methods suggested by Tsinghua University, Saskachewan, etc., as shown in Figure 12.9. In log-log plot all of the three relationships are approximately straight lines with slope ranging from -1.0 to -1.25.

Actually, the effect of pipe diameter on pressure slope J_m can be theoretically analysed based on Equations (12.6) to (12.8).

In hydrotransport practise some pipes are hydraulically smooth. Under such condition the pressure gradient J_m can be described by Equation 12.7. When the concentration of transported slurry is high and its τ_B is large, the second term in

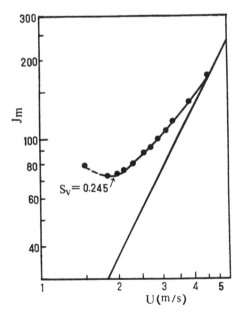

Figure 12.8. The relationship between J_m and U of suspension.

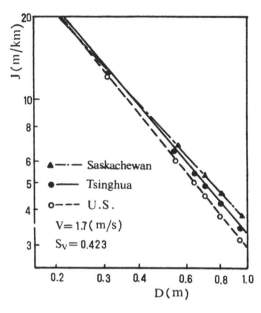

Figure 12.9. Relationship between diameter of pipe and energy gradient.

the bracket is much larger than 1. Therefore, the first term in the bracket can be neglected and in such case J_m is proportional to R^{-1} (that is, $(D/4)^{-1}$). When the yield stress τ_B of transported slurry is small and the velocity U is high, the second term in the bracket is much smaller than 1 and can be neglected. In such case J_m is proportional to $R^{-1.25}$ (that is, $(D/4)^{-1.25}$). Between these two extreme cases, J_m is proportional to -1-(-1.25) power of the diameter. For a fully developed turbulent flow the pressure gradient J_m can be described by Equation (12.8), in such case J_m is proportional to R^{-1} (that is, $(D/4)^{-1}$). In laminar flow region pressure gradient can be described by Equation (12.6). Hydrotransport is usually carried out at rather high velocity. Laminar flow must appear under the condition of extremely high yield stress τ_B. Therefore the first term in bracket is usually much smaller than the second one and can be neglected. Then the pressure gradient J_m will be proportional to R^{-1} (that is, $(D/4)^{-1}$).

As a summary, in all these cases pressure gradient J_m is proportional to -1-(-1.25) power of D. It is just the range shown in Figure 12.9.

12.5 OTHER EXAMPLES OF UTILIZATION OF HYPERCONCENTRATED FLOW

There are more fields in which the theory of hyperconcentrated flow can be applied. Drilling slurry in oil industry is a typical hyperconcentrated fluid. Fluidized bed is also a kind of hyperconcentrated flow. Newly developed dense coal slurry is hyperconcentrated fluid, too. The development and application of the theory of hyperconcentrated flow might favor solving some practical problems in these fields. And it is certain that the theory of hyperconcentrated flow might be found applicable in more fields.

REFERENCES

Charles, M.E. & R.A. Charles 1971. The use of heavy media in the pipeline transport of particulate solids. In I. Zandi (ed.), *Advances in solid-liquid flow in pipes and its application*: 187-197. Pergamon Press, 1977.

Elliott, D.E. & B.J. Gliddon 1970. Hydraulic transport of coal at high concentration. *Proc. of Hydrotransport* 1: G2-25.

Guo, Z., B. Zhou, L. Lin & D. Li 1985. The hyperconcentrated flow and its related problems in operation at Hengshan Reservoir. *Proc. of International Workshop on Flow at Hyperconcentrations of Sediment, Publication of IRTCES*.

Kenchington, J.M. 1978. Prediction of pressure gradient in dense phase conveying. *Proc. of Hydrotransport* 5: D7, 91-102.

Klose, R. & W.-D. Kunst 1985. Densecoal-Dense phase flow behaviour of Datong, Fugu and Shenmu coal and densecoal combustion. *Proc. of International Workshop on Flow at Hyperconcentrtations of Sediment, Publicatiion of IRTCES*.

Liujiaxia Hydroelectric Power Plant 1978. Releasing sediment in the form of density current in Liujiaxia Reservoir (in Chinese). *Selected Papers of the Symposium on Sediment Problems on the Yellow River*, Vol. 1-2: 1-8.

Luohui Research Group of Hyperconcentrated Irrigation/Warping 1975. Hyperconcentrated Irrigation/Warping in Luohui Irrigation district (in Chinese). *Shannxi Hydraulic Science and Technology*, 1975, No. 4:8-35.

Qi, P. 1989. Research on the guidance of training wandering river channels of the Lower Yellow River (in Chinese). Paper for Sino-US Workshop on Hyperconcentrated Flow and Debris Flow, (1989).

Qian, N. (Ning Chien), K. Wang, L. Yan & R. Fu 1980. The Sourse of coarse sediment in the Middle Reaches of the Yellow River and its effect on the siltation of the Lower Yelleow River (in Chinese). *Proc. of The International Symposium on River Sedimentation*, pp.53-62. Publication of IRTCES.

Ren, Z. & G. Zhao 1973. The preliminary experience of the operation of Heisonglin Reservoir – Storing clear water, releasing muddy water and hyperconcentrated irrigation/warping downstream (in Chinese). *Proc. of Reservoir Sedimentation*, pp. 169-183.

Shaanxi Research Group of Hyperconcentrated Irrigation/Warping 1976. Preliminary summary on hyperconcentrated irrigation/warping in Jinghui, Luohui and Baojixia Irrigation Districts (in Chinese). *Proc. of Sediment Research on the Yellow River*, Vol. 3: 107-137.

Wan, Z. & Y. Xu 1984. The utilization of hyperconcentrated flow and its mechanism. *Proc. of 4th Congress, Asian and Pacific Division, International Association for Hydraulic Research*.

Xu, Y. & Z. Wan 1985. The transport and utilization of hyperconcentrated flow in Luohui Irrigation District. *Proc. of International Workshop on Flow at Hyperconcentrations of Sediment*, Publication of IRTCES.

Xu, Y., Z. Wan, H. Shi & B. Jiang 1989. Utilizing sediment as resources in Luohui Irrigation

District. *Proc. of 4th International Symposium on River Sedimentation*, pp. 1618-1625.
Zhou, B., Z. Guo, L. Lin 1989. Regulating water and sediment for preserving storage capacity in Hengshan Reservoir. *Proc. of 4th International Symposium on River Sedimentation*, pp. 1205-1212.

Subject index

Printed and bound by CPI Group (UK) Ltd, Croydon, CR0 4YY

23/10/2024

01777686-0007